ROUTLEDGE LIBRARY EDITIONS:
ETHICS

Volume 39

THE RATIONAL FOUNDATIONS OF ETHICS

THE RATIONAL FOUNDATIONS OF ETHICS

T.L.S. SPRIGGE

LONDON AND NEW YORK

First published in 1988 by Routledge

This edition first published in 2021
by Routledge
2 Park Square, Milton Park, Abingdon, Oxon OX14 4RN

and by Routledge
52 Vanderbilt Avenue, New York, NY 10017

Routledge is an imprint of the Taylor & Francis Group, an informa business

British Library Cataloguing in Publication Data
A catalogue record for this book is available from the British Library

ISBN: 978-0-367-85624-3 (Set)
ISBN: 978-1-00-305260-9 (Set) (ebk)
ISBN: 978-0-367-50269-0 (Volume 39) (hbk)
ISBN: 978-1-00-304933-3 (Volume 39) (ebk)

The Rational Foundations of Ethics

T.L.S. Sprigge

London and New York

First published in 1988 by
Routledge & Kegan Paul Ltd

First published in paperback in 1990
by Routledge
11 New Fetter Lane, London EC4P 4EE

Simultaneously published in the USA and Canada
by Routledge
a division of Routledge, Chapman and Hall, Inc.
29 West 35th Street, New York, NY 10001

Printed in Great Britain by T J Press (Padstow) Ltd,
Padstow, Cornwall

British Library Cataloguing in Publication Data

Sprigge, T. L. S. (Timothy Laura Squire)
The rational foundations of ethics.
1. Moral philosophy
I. Title
170

Library of Congress Cataloging in Publication Data

Sprigge, Timothy L. S.
The rational foundations of ethics.

(The problems of philosophy)
Bibliography: p.
Includes index:.
1. Ethics. I. Title. II. Series: Problems of philosophy
BJ1012.S595 1988 171 .5 87–12894

ISBN 0-415-05519-9

To Giglia

Contents

Note

The following are the main moral philosophers whose views receive some discussion: Bentham, Mill, Sidgwick, Smart, Hare, Williams in chapter I; Moore and also Ross in chapter II; Ayer, Stevenson, and also Bambrough, Lovibond, Mackie, McDowell in chapter III; Spinoza, Hobbes, Hutcheson, Hume, Butler, Kant, Schopenhauer, Bradley, Royce, Santayana in chapter IV; Rawls and Dworkin in chapter VIII. The bibliography at the end gives sources for views ascribed to authors. The aim has been to pay due regard to conventional opinion as to which authors most require attention, while avoiding that rigid adherence to the current canon which makes each survey a repetition of the last.

The division into a Part One, offering a historical review, and a Part Two developing my own views, is in accordance with the design of the series to which the book belongs. However, some positions of my own, which seem most conveniently put as reflections on a historical author, are developed more fully in Part One. This applies particularly to discussion of G. E. Moore in Chapter II.

Acknowledgements

I thank Ted Honderich for having asked me to write this book, and for many discussions on ethical and related issues over the years. As this book is the fruit of reflections on ethics over many years I would also like to thank my earliest professional guides in this area, R.T.H. Redpath and A.J. Ayer. Among several others in discussion with whom I have advanced my understanding of many of the issues involved are my brother, Robert Sprigge, Bernard Harrison, Tony Grayling, Leemon McHenry, and Vinit Haksar, Margaret Paton, and other members of the philosophy department at Edinburgh.

Introduction

This work is concerned with three basic questions about moral judgements and other related value judgements.

(1) Do they possess objective truth or falsehood?
(2) Is there a rational method (or methods) for deciding whether or not to accept such a judgement? If so, what is it?
(3) Is there a rational method (or methods) for deciding whether or not to accept such a judgement, use of which will in the long run produce convergence among all those who use it (or them) efficiently?

These questions typically arise in ethics because many thinkers have thought that the answer to them reveals a striking difference between moral and other related value judgements and propositions typically called factual, such as scientific ones or some matter of fact statements of daily life. It is supposed that while our questions about moral and related value judgements may have to be answered negatively, those about factual propositions can be answered positively. However, some philosophers give a largely negative answer to one or more of these questions when they are raised about factual propositions, while an even greater number answer them in ways which make of truth and rationality themselves something seemingly so 'subjective' that the reasons often given for a contrastingly negative answer in the case of moral and value judgements largely fall.

Thus any particular view about the correct answer to these questions about moral and related value judgements will have a very different complexion according as to the sort of answer which is given or presupposed to corresponding questions about what are typically regarded as purely factual propositions. We could approach the questions much more easily if we could give examples of propositions susceptible of unproblematically objective truth and rational support of a kind calculated to produce convergence, and then consider how far moral and related value judgements are or are not radically different in kind. However, even if there are such propositions, there is little agreement among philosophers as to which they are or what a positive answer to such questions about them would mean. It follows that if I now give a brief explanation of what these questions mean, it can only be one which glosses over innumerable difficulties. However, some such attempt must be made in order to get us going.

With these cautions, I suggest that a judgement or proposition will be objectively true if and only if the activity of assenting to it is a way of registering the reality of something which is there independently (at least so far as any direct connection goes) of that activity.

An objectively true judgement or proposition (we must be clear) may concern something 'subjective' such as a state of mind. It is only required that it register a fact not directly dependent upon its being registered by assent to that proposition. That is true of the assertion that someone has a pain, and it is true of an assertion about someone's views concerning morality. If someone holds that moral and related value judgements can be objectively true or false, he need not be holding that these ethical truths would hold in the absence of all thinking. Obviously if there were no conscious individuals there could be no right or wrong actions (though some think a truth would still hold as to which would be right or wrong).

The remarks I now offer about the meaning of 'rational' must be treated with similar caution. It seems to have two main implications. First, a rational method must be such that there is an objective and accessible truth as to whether someone is applying it, so that one who uses it has a right to be confident that he is doing so. Second, it must have an appropriateness to the assertions on the acceptance or otherwise of which it is used to decide, in that it assists us, and can be seen to assist us, to obtain the ends for which such assertions are primarily accepted or rejected.

Traditionally, the two main rational methods for deciding whether to accept assertions (which go beyond mere observational or memory reports, however these are conceived) or not are deduction and induction.

By deduction one moves from assertions one already accepts to other assertions which one can recognize as absolutely bound to be true if the first are.

Induction is, roughly, moving from acceptance of the premisses that a certain phenomenon has very often occurred in the past in a certain relation to another, and that there are no cases of which one should know where it has occurred without being in such a relation to that other phenomenon, to the tentative belief that it only occurs in such a relation. Alternatively, it is implicit use of a generalisation, established in that way, to infer the existence of one particular state of affairs from another. 'Induction' also refers to inferences to generalisations about the proportion of cases in which certain phenomena are connected in general from the proportions present in observed cases, and for inferences which make use of such generalisations.

The rationality of induction is sometimes disputed. Rational or not, it needs to be supplemented by what C.S. Peirce called abduction, and what Karl Popper called the method of conjecture and refutation. Essentially, this consists in putting forward a conjectural explanation of some observed fact, or other supposed fact, and testing it by deducing consequences from it which we can observe, or otherwise determine, as holding or not.

Another type of rational method may be that of seeking for theories and accounts of reality which combine, to the maximal possible degree, both coherence and comprehensiveness. By 'coherence' is meant not mere logical consistency but co-membership in a system in which propositions inter-relate so as to provide each other with mutual support. The comprehensiveness of a theory is a matter of the wideness and variety of its implications. Theories which lack it risk winning other such virtues as those of consistency and coherence merely by neglecting facts which we should know about and remember. There are other methods, too, which are sometimes recommended as more or less basic to rationality, such as the dialectical method and the method of clear and distinct ideas, which would require attention in a more complete account.

Finally, let us note that some philosophers recognize a kind of rational insight into necessary truth as a rational method, or at least as an operation of reason, which should be considered alongside the other methods in considering the role of reason in ethics. One way of justifying the existence of such rational insight is that deduction – which is the most basic sort of rationality mentioned, for all the others presuppose it – itself presupposes an insight into necessary truths which affirm that if one thing is true, then so must another be. The typical examples of necessary truths supposed to be required in deduction are truths of a logical or formal kind, such as 'If all Xs are Ys, and this is an X, then this is a Y', but once grant the existence of rational insight, it is said, then you should be open to the possibility of rational insight into other sorts of necessary truth, of a less formal kind. Some have thought that truths which can only be known by a similar rational insight but which are not merely formal are basic to ethics. An example might be 'It is wrong to cause pain to others for your own enjoyment'.

We should not think of these various rational methods as rivals but as pervasive features of most rational enquiry. At any rate they are all supposed to give us guidance in the discovery of truth.

This may not be all they can do, however. Some of them may give us a good way of deciding whether to accept assertions in contexts where we are not necessarily seeking truth. Abduction may give us grounds for accepting a scientific hypothesis without making any strong claims for its truth. Some thinkers even hold that scientific hypotheses which concern the laws of nature are not the sorts of thing which can be true or false in any genuine way; rather are they recipes for arriving at true judgements.

Turning now to the potential of these methods for producing convergence, we may suppose at first that this is a necessary feature of a rational method. This seems especially so where we suppose ourselves to have a rational method for discovering truth. For surely truth is one, that is, consistent with itself. However, it is not really at all clear that even quite objective truth need be one in any relevant sense. Perhaps there are alternative and incompatible types of proposition, each rationally supportable, and each registering a reality which is independently there but which one can only take account of through remaining for the time being unaware of other realities. And even if truth is one, but the proper purpose of moral and related value judgements is not to arrive at it, there may be rational methods for arriving at one's moral opinions which carry no special promise of convergence on the part of all who use them.

In Part One of this work I consider a sample of philosophical theories of ethics which give or imply an answer to these questions. In Part Two I venture my own answer to them. I might note that there can be no question of an exhaustive coverage in Part One, and that there are bound to be different opinions as to which thinkers it was most important to cover, and how thoroughly. I have made my task more manageable by not considering any thinkers before the seventeenth century. Some important moral philosophers, including some contemporary ones, are wholly or largely ignored, because they are concerned less with the foundations of ethics than with direct ethical reasoning. That this is so of many contemporary moral philosophers represents what has mostly been a salutary shift from an earlier period when philosophers were almost exclusively concerned with foundations, and hardly at all with moral issues, but it is also important to turn on occasion to the more foundational issues which are the topic of this book. I regret, however, that limitations of space have not allowed me to give more attention than I have to the way in which the basic ethical position advocated here bears on concrete moral problems, both old and very new, and on political issues.

Philosophers are as bad at reaching agreement with each other on the nature of moral judgement (or anything else) as humans in general seem to be on what is right and wrong, or rather much worse. This is not the time to consider why that is so. However, the following suggestion should be born in mind in reading this or any other philosophical work. Philosophy may be preeminently a subject where insight into one aspect of things tends to blind us to other aspects, and we should try to draw upon other people's insights positively in order to advance our own halting and necessarily personal, but not therefore necessarily wrong, grasp on things.

PART ONE

CHAPTER I

Utilitarianism

1. We are concerned with three main questions in this work: Do moral judgements possess objective truth or falsehoood? Is there a rational means of testing their acceptability? Is this means calculated to bring congruence on ethical matters among those who use it properly?

A theory which, in its classic form, gives a pretty positive answer to at least the first two of these questions is utilitarianism. This is the doctrine – to put it roughly – that actions are right or wrong according as to whether they increase or decrease the amount of happiness in the world. Such judgements are supposed to be true or false in a fairly straightforward way, and to be testable by empirical enquiry or perhaps appeal to the common experience of the human race. In answer to the third question, the utilitarian might suggest that there is at least as great a tendency to convergence on these bases as pertains to any of the social sciences (which may not be saying much).

The utilitarian need not insist that these are the correct answers for all uses of ethical language or types of moral judgement. He may prefer only to claim that they apply once ethical language and judgement are understood in the way he recommends. Thus he may admit that words like 'right' and 'wrong' can be, and and often are, understood in a non-utilitarian fashion, perhaps simply to express ill considered emotions in statements incapable of rational testing or of being true or false. This, however, seems to him the condemnation of such uses of the words and such forms of moral judgement. As against the chaos, or alternatively the uncritical archaism, of such moral thought, he recommends that we tidy up our ethical language so that the words have a firm utilitarian sense. They are then used to support recommendations concerning the organization of society and personal conduct resting on straight matters of fact, or at least the best available opinion on matters of fact, of the one kind around which clear minded people can expect to rally as a basis for human cooperation. If the utilitarian looks at it in this way, he takes it as a criterion for an acceptable use of ethical words, and way of understanding moral judgement, that it should give them a factual content which is the only one which it is sensible to expect people in general to endorse as a sensible guide to acceptable conduct. Thus Jeremy Bentham, the great legal and social reformer, said that when interpreted according to the principle of utility 'the words *ought*, and *right* and *wrong*, and others of that stamp, have a meaning: when otherwise, they have none' (*Introduction to the Principles of Morals and Legislation*, Chapter One, §10).

Bentham (1748–1832) is properly regarded as the founder of utilitarianism. Although various thinkers before him had formulated much the same basic principle of utility as basic to ethics none had used it so systematically as the basis for rethinking all moral and social arrangements. It will be well, therefore, to start with his version of it.

According to this principle, as Bentham understands it, the ideal method for determining whether an individual's action, or a legislative enactment, is right or wrong would be through evaluation of its total tendency to promote happiness, on the one hand, and to promote unhappiness on the other; if the former predominates the action is right, if the latter it is wrong. To clarify this, Bentham lists seven so-called dimensions of pleasure and pain: (1) intensity; (2) duration; (3) certainty or uncertainty; (4) propinquity or remoteness; (5) fecundity; (6) purity; (7) extent.

Of these (1), (2) and (3) are those of prime importance. If we knew all the pleasures and pains liable to be produced by an action, and could assign a degree of intensity, a duration, and a probability (represented as a fraction of one) to each, then the multiplication of these by each other, treating pleasure as a positive and pain as a negative quantity, would give the total positive or negative value of the action's consequences, and it would be a good or right act if the result was positive, bad or wrong if it was negative.

Duration and probability, at least considered in the abstract, are relatively straightforward, but what is a pleasure's or pain's degree of intensity? The answer is that it is simply the degree of an experience's pleasantness or painfulness taken at a moment, or its average pleasantness or painfulness over a period. If we conceive of the least pleasant experience which is still a pleasure, then one way of quantifying its intensity would be by characterising the number of times more pleasant a pleasure is at a particular moment than that. Perhaps the slightest pleasure of which I can conceive is that of sucking a boiled sweet. Then listening to Beethoven's fifth symphony, well performed, and when one is in the mood to enjoy it, might be – let us say, on the average, taken moment by moment – one thousand times more pleasurable. Similarly, we should evaluate pains as so many times worse than a minimal pain.

Of the other dimensions, purity, in Bentham's special sense, means the chance of a pleasure not leading to pains, or a pain not leading to pleasures, while the fecundity is the extent to which a pleasure breeds or leads to other pleasures or a pain to other pains. The idea is that we may know that some pleasures or pains have a generally high or low degree of purity or fecundity without its being practicable on a particular occasion to specify and evaluate specifically the precise further pleasures or pains which are likely to ensue from them. Thus the pleasures of heroin can be dismissed summarily as counting against, rather than for, the action of taking heroin, if we can say that it is a highly impure pleasure (liable to lead to much wretchedness later). Extent is the number of people who will have the pleasures or pains, and it will be invoked when we are thinking in terms of some average effect on persons affected rather than of named individuals. As for propinquity it seems that he included this mainly because his theory of value was closely linked with his theory of motivation, and it is psychologically true that we are more influenced by thoughts of the more immediate than the more remote future. It seems better not to include it as a criterion of value, once it is clearly distinguished from probability.

We are not only to include pleasures and pains liable to be produced by an action in estimating its rightness or wrongness, but pleasures and pains liable to be prevented by it. Pains prevented, given a value in accordance with the same dimensions of intensity, duration, probability, extent count along with pleasures produced, in favour of an action, while pleasures prevented, count along with the pains produced, against it. Many have thought it an objectionable feature of utilitarianism, in its classic formulations, that pleasure and pain are supposedly set off against each other in this simple way. This seems particularly objectionable when pleasures and pains *prevented* come into account, for one may well think it worse in an action to produce pain than merely to hinder a corresponding amount of pleasure occurring. However, it is a problem to find an alternative approach. It will not do to adopt a so-called negative utilitarianism, once advocated by Karl Popper, for which all that matters is the prevention of pain. That seems to favour mass killing of all who are liable to feel pain at all.

It is no real objection to Bentham that we can only look for an approximation to the truth about these matters of pleasure and pain. It would be a more serious objection if there were no real truth there at all to which we can hope our judgements may approximate. There are at least two reasons for thinking this may be so. First, it appears that there may be no truth as to how many times more pleasant or painful one experience is than another. Second, there may be no one truth as to what is probable and to what extent. Moreover, there are difficulties in knowing how to relate different sorts of probabilities such as those which concern the degree of credit to be given to a judgement and those which are of a statistical nature. There are, however, some indications that Bentham did not think there was a precise truth to these matters, but that a certain free play left by the facts did not matter too much in practice.

2. Sometimes Bentham is represented as having held a significantly different view from that just described, according to which an action which it is right for me to do at any moment is either one which produces a greater surplus of pleasure over pain than any alternative action then open to me (in which case it is the one right action) or produces as great a surplus as any alternative (in which case it is *a* right action), while all actions not thus right are wrong. We may call this the *rigorist* interpretation or version of utilitarianism.

The requirement implied in this formulation, that I am always morally obliged to do the best I can, is accepted as reasonable by some commentators, while it is thought of as unreasonably demanding by others. (Samuel Scheffler has recently examined this matter in an especially interesting way.) Take, by way of example, a devoted nurse who is quite exceptionally sensitive to patients' needs. Her nursing activity would be right on our first less rigoristic formulation of the Benthamite view, for it does good by way of lessening pain and promoting pleasure and either does no harm at all, or none that is significant. However, on the second, more rigorist, view, her action would turn out to be wrong, should it be true that with even more effort she could have relieved still more suffering, or caused just a little more happiness (without countervailing harm). This implication of rigorism seems a bit absurd.

It is sometimes thought that rigorism can take adequate account of what seems troubling here by recognizing that if we ask too much of ourselves we damage our power to do good by becoming worn out or embittered. But it remains problematic whether the utilitarian should condemn as wrong all actions which are not the very best, or equal best. In favour of rigorism, there is the apparent reasonableness of saying that if one fails to do all one can to promote happiness (and, in particular, to reduce suffering) one has not done the best one could do, and that must be wrong. To allow people to comfort themselves that less than the best (or equal best) is good enough, is, it may be said, just a recipe for idle complacency. Against rigorism, is the sense that it is rather absurd to lump all who are not utilitarian saints with actual wrongdoers.

Actually, the debate between rigorism and our original less rigorist version of utilitarianism has a morally earnest character alien to Bentham. However, the issue is a lively one today, since some contemporary utilitarians adopt a highly rigorist outlook (arguing, for example, that not to have done all one personally might to feed the starving in far off places is as bad as murder) which other moral philosophers see as the reductio ad absurdum of utilitarianism. (See GLOVER and SINGER.) Since Bentham's position is usually assumed to be rigorist, it is worth pointing out that his main formulations, such as we have described above, seem to be non-rigorist. Undoubtedly, he would think an action the better the more it augmented happiness, but it seems that it is only wrong, for him, if it actually augments unhappiness.

However, one feature of his formulations does suggest the rigorist view. For we have seen that the bad consequences of an action are supposed to consist not just in pain, but also in loss of pleasure, and the good consequences of an action consist not just in pleasure, but in pain prevented. He is bound to say this, indeed, for if we considered only the positive consequences of an action, and not those which would have occurred if it had not been done, one could not count in the suffering which an action prevented in its favour, which would be absurd. However, once started on this track it may seem difficult to keep clear of rigorism, for it is unclear how the good one might have done but did not can be discounted from the things prevented by what one did instead.

It seems reasonable, however, to distinguish between two different types of thing which might or would have happened if the agent had not done an action. First, there are things which might or would have happened as consequences of some other action which he might have done instead. Secondly there are things which would have occurred, if he had not acted thus, but not as consequences of some alternative action of his. One can then say that it is only things of the second sort which count as things his action prevented, when one is calculating its good and bad effects, and thus distinguish Bentham's criterion of rightness and wrongness from a rigorist one. It stands as the view that an action is wrong if its consequences, in terms of pain promoted and pleasure prevented, outweigh its consequences in terms of pleasure promoted and pain prevented, and that other actions are right, even if not the best possible.

Yet this formulation still has an implication which Bentham would not really have accepted. It implies that if one achieves a desirable end (in terms, say, of pain prevented) but at a greater cost of suffering than was necessary for it, one's act was right, though not the best, even if one could have achieved that end by a less drastic means. This goes against Bentham's insistence that one must always use the means least costly in terms of distress caused. To meet this point I think one must say that Bentham's view was, in effect, that a right action must not only do more good than harm, *but must also be such that neither the particular good it does, nor any other comparable good which might have substituted for it, could have been achieved at less cost in terms of harm done.*

That utilitarianism needs some such additional clause to be in the intended spirit of Bentham is beyond doubt. However, the addition is almost as problematic as it is important. Consider the dispute over fox hunting and suppose it agreed that the pleasure of men, hounds and perhaps horses, outweigh the pain of the fox. Or consider, similarly, the cruelty of the Roman circus and suppose that there are enough happy spectators to outweigh the pain of the victims. Surely Bentham would want to say that they should have sought their pleasure in other ways which would, in fact, have been equivalent, but it is quite a problem to say what is meant by 'equivalent' here. Still, without returning on our tracks and urging the rigorist position (that one must always do as much good as possible), there must be some kind of requirement that one do as little harm as possible, in terms of actual pain (though not perhaps in terms of pleasure prevented) in order to achieve the good one does.

This formulation still allows it to be a justifiable ground for causing pain that it will promote a 'greater' pleasure, when no equivalent pleasure can be obtained otherwise. One may well object to this, but perhaps not on grounds which can be called Benthamite. However, a theory which was recognizably utilitarian in spirit might seek to give priority in some way to the prevention of suffering over that of the promotion of happiness.

3. Bentham combined his utilitarian ethical theory with a hedonistic psychological theory according to which each of us necessarily seeks to maximise his own happiness. This is usually thought a highly problematic combination. What point has an ethical directive to pursue the general happiness addressed to beings who will necessarily seek only their own?

This is not a problem for most contemporary utilitarians since they usually hold no such psychological theory. But whatever difficulties attach to Bentham's particular position, an ethical outlook which, like that of some more recent utilitarians, is not associated with some theory of human motivation is also somewhat unsatisfactory. Surely an ethical theory is not much use without some view as to what may induce humans to live by it. So it will be worth while considering briefly how Bentham saw the relation between ethics and psychology.

When someone does something we can ask (1) what he actually did; (2) what he intended to do; (3) what his motive was. What he actually did was right or wrong according to the criteria we have discussed, while his intention was right or wrong, on Bentham's view, according as to whether his action would have been right or wrong if things had turned out as he expected. His motive is the kind of pleasure for himself (or pain avoided) at which he was ultimately aiming. Bentham tries to show that on no useful classification of motives can one divide them into those which are always good or bad. Take, for example, the motive of becoming rich or winning admiration. This is neither good nor bad in itself, it is simply a normal bit of human psychology, operating more or less strongly in different people. To regard motives themselves as good or bad is idle; they are simply the raw material of human psychology with which the legislator or social engineer must deal. The important thing is to create a society in which the motives people actually have will operate so as to generate good intentions, such as will normally produce good actions, that is, ones which augment happiness.

But what part does the moralist, as opposed to the legislator or social engineer, play in Bentham's scheme? Curiously he comes out as another kind of social engineer who tries to show people that granted their fundamental motives (to obtain a range of pleasures and avoid a range of pains) they will do best to act rightly (in terms of the general happiness). But what if it is untrue that this is so? Then the moralist had better keep quiet on the matter, while turning to the social engineer to modify the social situation so that 'moral goodness' becomes 'the best policy'.

Thus many usual sorts of moral concern are simply absent from Bentham. People are what they are, but some acts are good, others bad, and legislators, reformers and 'moralists' should aim at organizing society so that human beings as they really are find themselves in circumstances where they tend to do the former. This obviously poses the problem: what motivates the moralists, social engineers, and legislators, who must be Bentham's main intended audience?

Bentham's views on this shifted. Originally he thought of his main audience as consisting in enlightened rulers who happened to have power and who with sufficient for themselves already, sought their main further pleasure in seeking the happiness of their subjects. His mature view is of more interest to us today. It was something like this. We virtually all take some pleasure in the pleasure of others and find pain in their suffering. Where we are not dealing solely with our own affairs but forming general preferences as to the kind of society we would like to live in, the main pleasures and pains we are concerned with are precisely these pleasures and pains of sympathy. Thus at that level we regard the social arrangements and the general customs of society, from the point of view of, so to speak, benevolent part time social engineers, that is in terms of their effects on the general happiness, and become a suitable audience to be energised towards social reform by Benthamite recommendations. As such, and also for most of us as being personally advantageous, we will put a high premium on democratic institutions.

4. The next great utilitarian thinker was J.S. Mill (1806–73), who attempted to improve on Bentham's utilitarianism in various ways.

Mill's most famous innovation lay in his claim that quality of pleasure counts as well as quantity. Suppose that one is choosing between two actions, both of which produce either no pain or the same amount, while one produces *more* pleasure. Bentham would say that the one which produces more pleasure is the better action, but Mill would say that it need not be, if the quality of the pleasure which is lesser in amount is sufficiently much higher in quality. (See his *Utilitarianism* chapter one.)

Mill does not discuss whether pains as well as pleasures differ in quality in an ethically significant way. It seems likely he would have said they did. He might, for example (having been somewhat high minded) have thought of the distress of a guilty conscience as worse than a quantitatively equal amount of physical pain. If so, an immense difference would arise in decisions on right and wrong between Bentham and Mill.

So it would seem, at least. Yet there is some difficulty in knowing whether Mill really meant to say anything very different from Bentham. If one takes 'amount' as referring to duration and extent (number of people affected), perhaps together with probability and proximity, Bentham certainly does not think that all that matters is amount, since this would be to forget intensity, which seems to mean simply the extent to which it is actually liked moment by moment. (It certainly does not mean any kind of degree of physical excitement.) Similarly, when Mill tries to explain how we can decide which is qualitatively the better of two pleasures, his actual answer is more or less that the higher pleasure is the one which is actually more liked by those in a position to compare them. So it is possible that the difference between Mill and Bentham here is merely apparent.

However, Mill is usually understood as holding that there may be two pleasures of which the less pleasant is the better in virtue of its higher quality, and that it is then morally more important to promote it than the other. Generations of philosophers have said that this is to abandon ethical hedonism, in a way which is quite inconsistent with Mill's avowed intentions. If one experience can be better than another without being pleasanter, does it not follow that there is something determining value besides pleasure? This seems a fairly devastating criticism.

But Mill may have meant something rather different, namely that, although all pleasures are, in themselves, and apart from their effects, good, and all pains, with a similar qualification, bad, they are pleasant and painful, and hence possessed of value or disvalue, in incommensurable ways. Thus in choosing between two alternative sets of pleasure one cannot necessarily decide which is preferable by an arithmetical calculation, nor could one necessarily do so even if one was omniscient.

Bentham's ideal calculation of the moral quality of an act presupposes that of any two pleasures the one is a definite number of times more pleasurable at any one moment than the other. Thus the average momentary pleasure, say, of hearing *Parsifal* on a certain occasion must be a definite number of times more or less pleasant than that of a particular episode of sucking a boiled sweet. This does seem rather ridiculous. Nor does it seem sufficient merely to qualify it by admitting a certain element of free play in the facts. This suggests that what Mill may have meant in saying that one pleasure is of higher quality than another is that it may be pleasanter without there being a quantifiable relation between them, in terms of which there must be some amount of the second which is as worth while, in hedonic terms, as the first.

The reasonableness of bringing some element of quality into even a strictly hedonistic system may be brought out by the following thought experiment. Suppose that we identify, or biological engineers construct, a slug like creature which, throughout its life, experiences a low level of pleasure consisting in an inarticulate feeling of humdrum comfort. Suppose also that it is somehow possible to place it in a space capsule in which it can survive, so far as can be seen, for ever, without any external aid. To construct this capsule is, however, immensely expensive. Suppose, next, that a catastrophe, which will wipe out life on this planet, threatens, and can only be prevented by an expenditure the same as would be needed for that capsule for the slug. A utilitarian dictator can either spend the resources he has at his disposal for developing the slug capsule or on saving life on this earth. He is a Benthamite and he reasons as follows. If I save life on this planet, it will only be saved for ten thousand years. (Let us take it that he is right in this.) If I let life on this planet die out and send the slug into space, then, since it will exist for ever, the quantity of pleasure it will enjoy will eventually outweigh all the pleasures enjoyed on this earth for ten thousand years.

This would be the only possible course for a strict Benthamite. Even if life on this planet were, for the next ten thousand years, to be pretty joyful, on the whole, the infinite number of low level moments of slug pleasure will be more valuable. If you do not like the idea of an infinite number you must still recognize that there must be some number of moments of slug pleasure which would outweigh any finite amount of joy on the earth.

So perhaps Mill's claim is that the pleasurableness of life is all that matters but that this cannot always be so well promoted by increasing the quantity of low level pleasure as by obtaining lesser amounts of high quality pleasure. Doubtless this means the end of straight arithmetical calculation, but it may still make sense to speak of informed judgements as to which patterns of pleasurable experience form more pleasurable lives.

5. The next major utilitarian thinker was Henry Sidgwick (1838–1900). His utilitarianism gains by being associated with a firm rejection of hedonistic psychology. However, matters are

complicated by his endorsement of the basic rationality, as opposed to psychological necessity, of egoism, which he sees as in conflict with the equal rationality of the concern with the general happiness required by utilitarian ethics.

A question which still greatly puzzles utilitarian thinkers, on which Sidgwick seems to have been the first to touch, is whether the utilitarian goal is the maximisation of total or of average welfare. (See BAYLES 2, and PARFIT.) Sidgwick himself settles without much ado for total welfare. (SIDGWICK p. 415.) (His approach is quantitative, for he firmly rejects Mill's problematic introduction of quality into the equation.) This has the implication that it is better to produce a sufficiently large population with low average happiness than a smaller one with high average happiness, since that way (with a sufficiently large population) the 'amount' of happiness will be larger. Indeed, so long as total pleasure just outweighs total pain an ever increasing population is likely to be a more effective way of increasing happiness than making life better for each individual. Many have thought this absurd and have concluded that the goal should be the increase of average happiness. However, this suggests that it might be a good thing to prevent anyone being born who would be less than maximally happy (on average, throughout his life) since he would be lowering the average happiness by his existence. This seems even more paradoxical.

There are genuine problems here as to how the utilitarian should view large scale government policies affecting population size. However, if our policies once allow for the possibility of the human race existing for ever, or into the indefinite future, one can take average happiness over all future time as the effective goal, even from the totalistic point of view, since the contribution to quantity of happiness from population size will anyway be maximal. (Cf. ATTFIELD p. 128.) Many attempts have been made to find a version of utilitarianism which avoids the oddities both of straight totalistic utilitarianism and straight average utilitarianism but none seems to be really satisfactory. (See especially PARFIT.)

Another question raised by Sidgwick still inspires lively debate. May it sometimes be better, from a utilitarian point of view, that people should not think as utilitarians? Perhaps it is best for the general happiness that people should believe that a certain obligation to keep promises holds independently of its effects on happiness? Sidgwick even floats the idea that the person who sees the truth of utilitarianism might, as a utilitarian, think it best to keep some aspects of his doctrine secret. Rather similar to this apparent endorsement of insincerity is the distinction he draws between qualities of character the praise of which is useful and qualities of character which are useful themselves. Only the first is praiseworthy for the utilitarian, since praise is essentially itself an act to be judged as good or harmful. Thus we should only praise conduct which needs to be stimulated in this way, even though other (perhaps more selfish) conduct which will be done anyway is itself good or better. For Sidgwick and for some modern utilitarians (such as J.J.C. Smart) recognition of this reduces conflict between utilitarianism and ordinary moral common sense, but for others it is a symptom of a kind of bad faith which is endemic to utilitarianism.

Finally, it is to be noted that Sidgwick holds that a kind of intuition, not capable of empirical or discursive proof, is needed to support utilitarianism. The statement that the actions we ought to do are those which promote the general happiness would be useless if it meant nothing more than that they do promote it, and that something more can only be given by an intuition. (Sidgwick thus avoids the naturalistic fallacy to be considered in the next chapter.)

6. Appeal to intuition is uncongenial to most utilitarians. We can find the beginnings of an alternative defence of utilitarianism, not turning merely on a question begging appeal to the meaning of ethical words, in Bentham. He contends, in effect, that ethical statements do indeed do something more or different than merely state facts, even facts about the effects of actions on the general happiness, for they pick out actions for approval or disapproval. Nonetheless, guidance of our approval and disapproval by the principle of utility is the only serious option for those who wish to base them on ascertainable facts to which we can all come to attach the same kind of importance. (Mill is much less clear on this.)

Modern utilitarians who are disinclined to appeal to intuition sometimes adopt an attitudinist view of ethical judgement, of the sort we shall examine in our third chapter. According to this, ethical propositions do not say something which can be either true or false in the sense of reporting or misreporting how things really are in some realm of values (though we can call them 'true' or 'false' to signal our agreement or disagreement). Rather, they express some 'attitude' of the speaker, such as some ultimate preference about how we should behave or society be organized, and invite the hearer to adopt it too.

A utilitarian who holds this 'attitudinist' view will see utilitarianism as the expression of a fundamental attitude in favour of a certain way of reaching decisions. He will not suppose that there is any possibility of proving that the principle of utility is *true* in any strong sense, but will put it forward as a guide to life which he endorses and, as a philosopher, seeks to make precise and free of confusion. He may think also that it has a basic appeal to human nature, which means most others will accept it too once their minds are cleared of confusion and superstition. A philosopher who has advocated utilitarianism in this spirit is J.J.C. Smart.

A utilitarian of this sort holds that once we have adopted a fundamentally utilitarian standpoint all specific ethical questions become factual and can be approached empirically. However, if people challenge the basic principle of utility one cannot say that they are mistaken in any kind of factual way, or that they are necessarily irrational. It may simply be a fundamental divergence of attitude.

7. A form of attitudinism which has been worked out with especial thoroughness is the prescriptivism of R.M. Hare. For him, the essential meaning of the key ethical terms lies in two features, universalisability and prescriptiveness. They are prescriptive in the sense that their role is to prescribe (which means much the same as recommend) some particular sort of conduct. Thus ethical statements cannot be true or false in a basic way, any more than a command can be. However, ethical words are also universalisable. It is linguistically unacceptable to apply them except on grounds which would always lead to further application of them in all similar cases. It follows that I am not using language properly if I say that you ought to do something, unless I hold by some universal principle from which this prescription follows and all other implications of which I would be prepared

to endorse, most notably those which would prescribe under certain circumstances that I do something.

When he first advocated these ideas Hare largely went along with the view of most attitudinists that a correct account of the meaning of ethical language has no definite implications as to what moral views one should take. But in successive writings Hare has gradually worked around to seeing the account of the very meaning of ethical language which he regards as correct as offering a kind of proof of utilitariansm. (See HARE 1,2,3.)

For, so he elaborately argues, one cannot, if one really thinks it through, accept the total set of prescriptions implied by any universal ethical rules except those which attach weight, in proportion to their strength, to the desirability of satisfying the desires which everyone affected by an action would have if they possessed proper prudence. This is because I can always ask 'Would I still prescribe that action if it was I who had that desire?' The idea is that one does not really accept a universal rule unless an imaginative attempt to put oneself in the place of everyone affected still leaves one happy with it, for only so does one accept the prescription that the action should be done in that case in which it would be oneself rather than the other who was in such a place.

8. The upshot is a version of what is known as preference utilitarianism, for which what counts in favour of an act is not that it promotes a kind of experience known as pleasure or prevents a kind of experience called pain, but that it provides people with what they would prefer to have and prevents their having what they would prefer not to have. Such considerations as the following are widely thought to favour preference over the older hedonistic utilitarianism.

(1) The fact that someone prefers X to Y, as shown in their behaviour, is empirically ascertainable in a way in which the greater intensity or duration, or even very existence, of a private feeling of pleasure or pain is not. It is also supposed that comparison between welfare gained or lost for different people is, though still difficult, less intractably so, if what is in question are not qualities of feeling but the satisfactions of behaviourally manifested preferences. (2) Then again it is regarded as doubtful whether there really is a specific quality of pleasure, and even of

pain, which covers all experiences that people either want to have or avoid, and it is thought obviously desirable that people should have experiences they want rather than those with a certain quality. (3) Even if the nature of these qualities of pleasure and pain are themselves unproblematic, it is said that what people want (or want to avoid) is not necessarily any kind of private experience, but just as often objective states of affairs in the public world, and it is thought no less reasonable to take account of wants of this sort than those for private experiences.

To me these supposed advantages of preference utilitarianism seem spurious.

(1) Doubtless a person's behaviour in choosing between alternatives is observable in a way in which his felt satisfaction or dissatisfaction is not. However, if preferences are understood merely as patterns of physical behaviour which tend towards certain results, then there would be no more moral reason for satisfying them than for assisting a computer carry out its programme when this was something undesired by any conscious being. Only if it is taken for granted that the preference behaviour is that of a conscious subject, does it, of itself, provide a reason for promoting the preferred end,– it would not matter in the least if there was no conscious individual there to mind about anything. But once granted that we have adequate ground for interpreting people's preferences as those of conscious beings, it is not clear why it should be denied that we have adequate clues as to whether they have pleasant or unpleasant experiences under various circumstances and for some judgements about the degree of the pleasure or pain involved which allow interpersonal comparisons.

(2) The denial that there is some one and the same quality of pleasantness and another of painfulness which mark all those experiences which it is desirable to promote or prevent is a more forceful objection to Benthamite utilitarianism. Preference utilitarianism is right to try to get away from this idea. It risks, however, giving the impression that what matters is not the quality of life as we each experience it in our individual consciousness, but simply what takes place in some objective world – which seems a betrayal of the whole point of utilitarianism. Perhaps Mill's qualitative utilitarianism points towards a better alternative to the view that pleasure and pain

are simply some kind of uniform sensation of which we want, respectively, as much and as little as possible. We can surely think of pleasure and pain as referring to felt qualities of experience without denying that these qualities are of radically different kinds.

(3) The third consideration supposedly favouring preference utilitarianism was that we may desire things other than experiences for ourselves. Once we reject psychological egoistic hedonism we have virtually admitted this fact. I may prefer that people should not be malicious about me behind my back, even if I am not to know or even be affected by it, and that certain deathbed wishes of mine be carried out without supposing I will persist to be affected by them. Moreover, I have preferences regarding the happiness of others which are not concerned with them merely as means to certain feelings for myself.

With such facts in mind, the preference utilitarian may suggest that our aim should be not just that people should somehow have as much subjective experience as possible of the kinds they most prefer, but that as much as possible of what they would like to have happen should happen. But this suggestion is full of difficulties. It seems to imply, for one thing, that if someone has very strong preferences about what happens beyond his own person he thereby renders it important that certain things be done or left undone which have little or nothing to do with his personal life. Indeed, this influence can extend beyond the grave, when he will not even know (let us assume) whether things are working out as he preferred.

9. Another variant of utilitarianism which commanded a good deal of support in fairly recent times was *rule* as opposed to *act* utilitarianism. (See TOULMIN, LYONS and BAYLES 1.) For the traditional 'act' utilitarianism *acts* are right because they maximise (or at least increase) welfare, while for rule utilitarianism *rules* are morally binding because general adherence to them maximises, or would maximise, welfare, individual acts being right or wrong in virtue of their conformity to such rules. Rules are, indeed, important for act utilitarianism, too, and not only as rules of thumb, but rather as felicific habits; still, the effects on people's inclination to stick with generally felicific rules is simply a consequence of an individual act to be weighed along with all of

its other consequences. For rule utilitarianism, in contrast, once a rule is shown to be felicific, it is established as something to be obeyed, unless perhaps in very special cases, and is not to be considered merely as one factor to be weighed against others. Thus granted institution of a rule against murder is in general felicific, we should abide by it even when individual calculation of the results of a particular murder might show it to be beneficent. Rule utilitarianism is often advocated, on grounds of fairness, as necessary to counter the excuses of the 'free rider' who does not bother to stick to a moral rule because enough others are doing so to produce its benefits.

The usual objection to rule utilitarianism is that if the sole point of a rule is to promote happiness it seems only sensible to jettison it when more happiness is gained thereby. On the face of it, this criticism carries the day against any rule utilitarianism which is genuinely distinguishable from act utilitarianism. The issue is complicated, however, by the variety of forms which rule utilitarianism can take. [LYONS, SCHEFFLER.]

10. Utilitarianism has been both vigorously defended and attacked in the last few decades. Many think that in spite of strenuous efforts by Mill, utilitarianism cannot really do justice to the concept of justice. Another important criticism, associated especially with Bernard Williams, is that utilitarianism ignores the real significance which life has for mature human beings.

For each of us (it is said) our lives are given sense by certain projects, whether it be that of being a certain sort of person, or of achieving something in such fields as the political, or cultural, or in our personal relations. For the person with these projects the forwarding of them has an importance to which utilitarianism cannot do justice, for it must regard them as simply among the many preferences of which as many as possible are to be satisfied. In a sense the real utilitarian only has one project he takes seriously, the satisfaction of preferences or desires (whatever they are for) or, in the older version, the maximisation of happiness. Yet if one takes what one does with one's life seriously, one cannot simply throw one's own most serious projects into the melting pot of desires and preferences along with everyone else's.

Suppose that I am committed to an ideal of conserving areas of natural beauty or variegated wild life in my country. If I am serious, I cannot simply see the preservation of the countryside as a desire of a few people like myself, to be weighed in the balance against the desires of others for holidays abroad, when it comes to deciding on whether a new airport should be built in some place of outstanding beauty.

A similar objection to utilitarianism urged by Williams, Alasdair MacIntyre and others is that utilitarianism goes with a manipulative approach to human life. It is said to be the philosophy of government administrators, dangerous in their hands, corrupting treated as the basis of private morality. Then it is also objected that utilitarian thinking, which has reached its apotheosis in modern cost benefit analysis, regards all values as commensurable, and therefore thinks of every harm as something which can be compensated for, reaching, it is felt, particularly repellent extremes when the value of a human life is calculated as something to be set against the goods achieved by a motorway or by economy in safety precautions at a factory.

These objections deserve to be taken seriously. However, one will do well to reflect on the following remark of Bentham (of which some of the second part of this work might be seen as an endorsement).

> When a man attempts to combat the principle of utility, it is with reasons drawn, without his being aware of it, from that very principle itself. His arguments, if they prove anything, prove not that the principle is *wrong*, but that, according to the applications he supposes to be made of it, it is *misapplied*.
> *Principles of Morals and Legislation* Chapter 1.

CHAPTER II

Intuitionism: Moore's Ideal Utilitarianism and Ross's Theory of Duty

1. For a long time after its publication in 1903 G.Moore's *Principia Ethica* was the most influential work of moral philosophy in English. More recent moral philosophy exhibits a certain boredom with Moore's approach. Nonetheless, it is an outstanding work and especially relevant to the present enquiry.

For Moore the 'science' of ethics is concerned with the property or quality *good* – with what it is, with what possesses it, and with how to produce as much of what possesses it as possible. (He allows from time to time that it is also concerned with the quality, *bad*, but *bad* takes a decidedly second place in his discussions.)

Moore contended that most moralists have confused three questions.

(1) What is the quality of good? What is that quality which good things have in common?
(2) What things, or main sorts of things, possess this property?
(3) What are the most effective means of promoting *the good*, that is, the stock of things which possess the *quality* good? Since goodness is a matter of degree, the extent to which one has promoted the good is a function both of the number of good things produced or things made good *and* of the degree of goodness to which they are raised.

Let us begin with his efforts to clear up the confusion of questions 2 and 3.

Things are often called good on account of their good effects.

It is an old theme that this runs to an infinite regress unless some things are 'good in themselves'. Thus the sense of 'good' which is fundamental for ethics is that in which it can be expanded to 'good in itself'.

We may observe in passing that while it seems obvious enough in the abstract that we should distinguish between what is good in itself and what is good as a means, more is at stake here than conceptual clarity. Rather few human activities seem to have their main point merely in themselves. This is true, in particular, of most of what we call *work*. The activities which seem most obviously to have their point in themselves are what we regard as typically leisure activities – art, playing games, joking. Thus there is a tendency, if we once accept a sharp distinction between good as means and good as end, to see the point of life in those leisure activities, or at least in activities which stand apart from the main work of the world, such as high culture and very private personal relations. Moore's own views, as we shall see, are fully in line with this.

Now while it may be difficult to gainsay the claim that these are the things on which it would be most worth spending time if there was ever no further need to struggle for the basics of human survival and comfort, life seems to be trivialised if culture and personal relations are held up as the be all and end all of what really matters. Some wariness is in order, then, about accepting this distinction as no more than uncontentious conceptual clarification. This is a point to which I shall be returning.

The tendency of moral philosophers, thinks Moore, is to rush into the third question, without having dealt with the first two, and without distinguishing the special nature of each. Ethics is normally thought of as dealing with questions of action, asking how we *ought* to act, what it is, in various circumstances, our *duty* to do, and how we can know it. As Moore sees it, questions as to what we ought to do are questions as to which actions are good, indeed best, as means. He thinks that we will never deal intelligently with these questions if we confuse them with questions about what is good in itself.

2. Let us now turn to Moore's first question: What is the property or quality which the word 'good', taken in its most fundamental sense, denotes? Moore's view is that it is a simple and therefore

indefinable property. Some things have it, and others lack it, and the things which have it have it in different degrees. (Note, however, that 'better' covers less bad, or being good as opposed to being bad, as well as degree of goodness.)

Moore insists that the statement that *good* is indefinable is not about the word 'good' but about the notion or object it denotes. Moreover, the word 'notion' is not meant to suggest that he is talking of an idea in our minds; he is concerned rather with that actual feature of good things which is picked out by the word 'good'. However, subsequent thinkers have often seen his claim as in effect about the meaning of the word 'good' and Moore himself at times expresses himself thus. Perhaps the conflict is not as great as all that. Moore would surely grant that there is an indefinability of the word which follows from the fact that what it labels is indefinable, while those who treat it as a statement about a word see it as turning upon what the word is supposed to stand for. Thus it is often convenient to treat it as the claim that no complex synonym can be provided which brings out the meaning of 'good' as say 'parent's brother' brings out the meaning of 'uncle'.

In saying that *good* is indefinable Moore compares it with *yellow*. He says that both are simple notions or objects without parts, contrasting them in this with the object *horse* which has parts, (head, hooves, tail etc) which can be mentioned in explaining the nature of this object.

This is rather confusing. The point is surely that the properties of being good and being yellow are not complex properties as being a horse is, which is (we may take him to be supposing) a matter of having a whole lot of simpler properties, such as having a certain sort of head, a certain sort of tail, and so forth. It seems misleading to express this in terms of a universal object horse which, like individual horses, has such parts as a head, a tail, and so forth.

Moore's way of putting it might be defended on the basis of a realist view of universals for which individual horses are horses because they participate in a universal object *horse*, and do so in virtue of the fact that they have parts participating in universal objects which are parts of the universal object horse and related to each other in ways which participate in the universal relations linking part to whole in the universal object. In contrast, an object would participate in the universal object goodness or yellowness without having to have parts which participate in some universal parts of that universal object.

In terms of this position Moore could say that the object or universal *horse* is made up of parts to which the parts of an individual horse correspond, while the object or universal *good* is not made up of parts to which the parts of an individual good thing correspond. It is thus something like the universal *yellow* which pervades a yellow object without requiring any particular arrangement of individual parts. (For Moore yellow is simply the quality we immediately apprehend in sight, not its physical basis.) There are, however, ways in which the analogy breaks down, for while a yellow object is yellow in every visible part, a good object is not necessarily good in every discernible part.

Moore might well have presented his claim that good, in contrast to *horse* and various other universals, is simple, as the statement that it is a quality, for there is a sense in which all the universals most naturally called qualities are simple in contrast to other universals, such as forms (e.g. shapes) or structures (e.g.the way in which an army is organized).

3. Moore has a famous argument in favour of the simplicity and indefinability of good. Consider any complex property which someone is tempted to identify with good. Does not this inclination arise from his believing in the goodness, the worth, of what has this property and from his belief that its goodness is an important ethical truth? However, if good and this property are simply the same thing, then the judgement that what has this property is good is just a tautology, and cannot serve as the kind of important value judgement it is meant to be.

It has sometimes been claimed, for example, that 'good' means *tending to promote the survival of the human species*. Evidently those who make the suggestion hold that things which promote the survival of the species are good, and that no others are, and that it is important to recognize this. But this could not be so if calling these things 'good' were just to reiterate that they promote the survival of the species. No one thinks it an important proposition that what promotes survival of the human species promotes survival of the human species. But those who offer to define 'good' in this way do think it important to insist that what promotes this survival is good. Thus they are implicitly recognizing that 'good' cannot just mean *promoting human survival*. How could it be of ethical significance that the word 'good', considered just as a sound, is used in a certain way? Ascribing goodness to this tendency must be saying something further about it.

It has been suggested that Moore's argument, if just, would count against all definition in which an expression's meaning is explained by some more complex expression [FRANKENA]. But that is not so. You can define 'uncle' as meaning *parent's brother* precisely because you recognize that it is not an important truth that someone's uncle is always his parent's brother. Its only interest lies in elucidating English usage. Moore is not denying that such elucidation is sometimes needed and possible, but bringing home to us that this is never what is going on when all things with a certain complex property are said to be good. Such claims always have a point which goes well beyond instruction in English usage.

Moore says that those who try to identify good with some complex property are committing what he calls the naturalistic fallacy. He so calls it because the properties with which people committing this fallacy typically identify good are natural properties. It is, however, committed equally by someone who identifies good with a supernatural property, such as tending to save the soul. (That it is good to save the soul, if it is, is not just a tautology, but an ethical judgement.)

The fallacy is committed equally if we identify good with some simple natural property, with pleasure, for example. If everything which is pleasurable is good that is a significant ethical truth, not a statement about the meaning of a word, or the tautology that whatever is pleasant is pleasant. Moore's account of what he means by a natural property is none too clear, but in effect it means something like *detectable by the senses or by scientific instruments*. He claims that it is never a mere tautology that what has some such property is good, but always a substantial claim in ethics. This can be established *ad hominem* to anyone who makes such an identification, for it is always done by people who attach importance to the claim that what has the natural property in question is good.

Is the fallacy bound to be present whenever anyone says

anything at all of the sort 'good is . . .', meaning to offer a definition? Moore's view seems to be that the only non-fallacious cases are those where we are given merely some obvious synonym for the most basic sense of 'good', like 'intrinsically valuable'. But he does come rather near making the rather empty point that it must be wrong to identify good with anything different from itself. His reply would be perhaps that the proof of the pudding is in the particular cases. Wherever it is thought important to claim that things of a certain sort are good, being of that sort cannot be what 'good' means (what the property *good* is). That permits rather trivial definitions of good such as 'good is what is intrinsically valuable', but rules out all more interesting definitions.

Sometimes Moore identifies good with *ought to exist for its own sake*. This rather dents the claim that good is completely simple. However, it does not affect the force of his reasons for saying that all definitions of good in natural or metaphysical terms are wrong headed.

4. In *Principia Ethica* (p. 41) Moore implies that what makes *good* a *non-natural* quality is that it could not exist by itself in time, as other properties could. (See also p. 124.) The idea is that properties of an object, such as the shape and colour of a thing, could, in principle, exist on their own ('in time' – as opposed to in some more Platonic way) apart from anything else, whereas *good* does not have the same possibility of existing as an independent object. Moore himself said later that there was little to be said in favour of this account. (SCHILPP p. 582.) The best account of what Moore thought so special about good is in an article called 'The Conception of Intrinsic Value' in his *Philosophical Studies*. Although he does not speak there of properties as being natural or not, his account of the sense in which for him there is such a thing as intrinsic value does, I think, serve as an account of what he was getting at in regarding *good* as non-natural.

He says there that the intrinsic value, (such as the *goodness* or *badness*) of an object is peculiar in that (a) it *depends* on the intrinsic nature of the object, while (b) it is not one of the object's intrinsic properties. His rather tortuous explanation seems to come to this. The intrinsic nature of a thing is the whole detailed character which it would share with anything else exactly like it, so far as that character is entirely a matter of what the thing is like, so to speak, within its own boundaries, while its intrinsic properties are the various discriminable elements of its intrinsic nature. Thus the intrinsic value of something follows from its intrinsic nature, so that nothing exactly like it could be of a different intrinsic value, while yet it is not one of the features one could properly list in indicating what something else would have to be like to be just like this thing.

Thus yellow and roundness might be intrinsic properties of a yellow ball as being among the properties in virtue of the possession of which another ball might be exactly like it.* The intrinsic value of the ball, in contrast – say its beauty – would follow from the intrinsic nature made up of such properties but not itself be part of that in virtue of which two balls might be exactly alike. (As examples of properties which are neither intrinsic properties nor intrinsic values, consider those of being in a certain house or evoking pleasure in an observer.)

As an illustration, consider a state of mind, which Moore would certainly think intrinsically bad, consisting in some precise form of enjoyment of a specific piece of sadistic pornography. Moore's view would be, I take it, that someone might know exactly what that state of mind was like, either by personal experience or imagination, without recognizing that it was bad, as he could not without taking in all those properties by sharing which another state of mind would be just like it. Its badness, therefore, would not belong to its intrinsic nature and not be one of its intrinsic properties. Yet he would hold that there could not be another state of mind just like it, without this being bad in just the same way and degree. So its badness would follow necessarily from its intrinsic nature.

An example of something similarly intrinsically good might be a feeling of deep awe at the sight of a magnificent cathedral. Anyone who imagined that precise experience would know its full intrinsic nature but they might or might not realize that this was a good experience.

These examples should bring home the great difference that

* To get the flavour of Moore's approach, you must think of a physical thing in the naivest way as having colour and shape in a manner having nothing to do with any actual or possible observer.

there is for Moore between intrinsic good and bad and anything of the nature of pleasure or pain. One could not know what an experience was like without knowing whether it was pleasurable or not, but one could know what it was like without recognizing its value, even though this follows necessarily from its being like just what it was.

I suggest, then, that the alleged non-naturalness of *good* and *bad* consists in their being properties of things which follow from what the things are like but which do not contribute to things being like what they are like. It is an important implication of this that they are never a matter of how the things which have them are related to other things. (In speaking of what a thing is like in this sense, one is not referring to its relation of resemblance to other things, only of that in it which could be the basis of such resemblance.)

Moore's terminology is complicated by the fact that he allows the possibility of properties which are not natural, without being non-natural in the special sense in which value properties are. For he distinguishes natural objects and properties – components of physical nature and the minds of organisms existing in it – from some present allegedly only in metaphysical realms postulated by certain philosophers. In these metaphysical realms things would have intrinsic natures and relational properties which would contrast with their intrinsic values just as in nature. Examples might be the distinguishing features of God or of so-called noumenal selves. The naturalistic fallacy is committed by the equation of intrinsic value with intrinsic properties and relational properties at the metaphysical, as much as at the natural, level.

5. The fact that, for Moore, the value of a thing follows necessarily from its intrinsic nature, from what it is like, makes it a little misleading to say, as is often done, that it is supposed to be always an *open question* whether something characterised in terms of its natural, or metaphysical, properties is good or not, and that this is his chief reason for regarding good as indefinable. That is too suggestive of positions for which there is no kind of necessary connection between a thing's natural character and its value. For Moore, in contrast, the intrinsic nature of a thing settles, without anyone having any choice in the matter, what

value a thing possesses. (Of course, 'things' here includes preeminently human experiences.)

Moore suggests that recognition of this may open our minds to the possibility that there is a plurality of goods. So long as we think that good must be identical with some one natural property we cannot but suppose that all good things have some such property in common. This explains why so many philosophers have supported theoretical hedonism. If one picks out some of the chief things which are good, one may find that the one thing they seem to have in common is that they are pleasurable experiences. Since one thinks that there must be something in common to all good things one may then conclude that it must be pleasure. But if one rids oneself of the idea that there need be any thing in common to good things, other than that they are good, one will be ready to recognize that there may be other good things which do not involve pleasure, that some things which involve pleasure may be good rather than bad, and that relative goodness need not be proportional to relative pleasurableness. It is not, indeed, ruled out by the logic of the naturalistic fallacy that degree of goodness and degree of pleasurableness might coincide, it is just that once one is free of the fallacy one will no longer see any reason to hold this.

Moore contends that if, having freed ourselves from the naturalistic fallacy, we ask what are the chief good things known to us, we will conclude that they are personal affection and the enjoyment of beautiful objects. They are not the only good things, but they are much the best, at least among good things of which we know. It is of the very nature of the case that there can be no proof of this. It is simply what presents itself to Moore as the truth when he reflects as carefully as he can, and what he therefore expects to present itself likewise to others.

6. Still, Moore does have a kind of ethical methodology. It has two main features. First, he holds that we should recognize that valuable things may be organic unities; second, he recommends a method of isolation.

According to the principle of organic unities, the value of a whole does not have to be the same as the sum of the values of its parts. Where it is different, Moore calls the whole in question an organic unity. Thus if we have a whole W, made up of parts X and Y then we cannot assume that the value of W is the value of X added to the value of Y, for W may be an organic unity. After all, no one would think that the beauty of an object is simply the sum of the beauties of its parts. If it were, breaking a beautiful statue would be harmless provided one kept the parts.

Moore's idea is that if W, which is made up of X and Y, exists, then the distinct values of X and Y, exist, so that the values of X and Y are certainly won to the universe, but that there is also another value, that of W as a whole, which could be positive or negative. Then the value *of* the whole W (the value it has, as Moore puts it, *on* the whole) consists of the value of W *as* a whole plus the value of X plus the value of Y. According as to whether the value of W as a whole is positive, negative or nil the value of W, on the whole, may be greater, less, or the same as the value of the sum of its parts. (When I speak of the value of the whole, I mean what Moore calls its value on the whole, that is the values of its parts plus its value as a whole.)

There is nothing very mysterious about this. For a whole, W, to exist it is not normally enough merely that its parts exist, rather it is required that they exist in a certain arrangement, in a certain set of relations to each other. Thus when W, composed of parts X and Y, exists, something exists besides X and Y, and it is not surprising that it should have its own fresh value.

However, the matter is more complicated than Moore acknowledges. For, after all, there is some difficulty in knowing what is meant by the sum of the values of the parts of something.

What, after all, do we mean by the parts of a thing? A thing could be regarded as made up of parts in different ways, according to how we divide it mentally. Is it certain that if we divide it in alternative ways, and attach a value to the parts in each case, the sum will always be the same? If not, then the notion of the sum of the value of the parts is not clear cut, and it is similarly not clear cut what is involved in the value of the whole not being the same as this sum. England, Scotland and Wales, and each of their counties or regions, are all parts of Britain, but it would be odd, in summing up the values of the parts of Britain, to include both the values of Scotland, and that of Lothian region.

However that may be, let us look at Moore's principle at work in one of his examples. He says that the consciousness of a beautiful object is a thing of great intrinsic value while the sum of the value of the two great parts which make it up – (1) consciousness (2) the beautiful object – is not very great (*Principia Ethica* p. 28). Moore does, indeed, somewhat notoriously, think that a beautiful object has some value in its own right, apart from consciousness of it, but he believes that this is slight as is also the value of mere consciousness apart from its objects. (One may well hold that such mere consciousness is an impossibility, but Moore treats it as making sense.)

But is the consciousness of a beautiful object properly conceived as consisting in two distinct things, consciousness and the object, standing in a certain relation (of the one being directed at the other) which makes no difference to what each of them is in itself? Does not their coming into this relation modify or add to the character of either of them? Can, for example, consciousness avoid being somehow enriched by a pervasive quality, which could not possibly have occurred except as a quality of just such a state of consciousness, when it becomes consciousness of a beautiful object, instead of consciousness of something else, or 'mere consciousness'? If so, may not this quality have a value of its own needing to be counted in when summing the value of the parts and may it not be this value (rather than the value of the whole as a whole) which adds to the value of the whole? Until such questions are satisfactorily answered, evaluation of the principle of organic unities remains problematic.

My point is not that we might count the relations between parts as themselves parts – though this is not necessarily mistaken* – but that the relations into which the parts enter in making up the whole affect their character and value so that they do not necessarily have the same value as they would have had out of that whole. In that case, the sum of the values of the parts of a thing should not be equated with the sum of the values they would have had outside that whole. From this it would follow that the principle of organic unities has no clear meaning.

* The main ground for rejecting the idea that the relations between the parts cannot themselves be parts of a thing, namely that relations are universals, is hardly available for the Moore of *Principia Ethica*, for whom a thing's properties are among its parts.

Moore, indeed, dismisses as incoherent the idea that the parts of a thing may have characteristics they could not have had apart from that whole. He notes that a whole with parts of which this was true would be an organic unity in a sense different from his, a sense, however, which he thinks nonsensical. I shall speak of the organic unities which Moore dismisses as nonsensical as 'organic wholes' to distinguish them from 'organic unities' in Moore's sense.

The reasons Moore gives for thinking that there cannot be organic wholes are not very compelling.** So unless we are entirely convinced that there are no such organic wholes, we should remain open to the possibility that some of Moore's organic unities are really organic wholes. They would perhaps be organic unities too in a sense close to Moore's but the way in which the value of the whole was not composed of the values of the parts will be different. Since their parts (or at least some of them) can only exist as parts of just such a whole, no definite sense can be attached to a contrast between the sum of the values of their parts and their value on the whole.

Consider someone's total state of consciousness at a particular moment, comprising various perceptual experiences, thoughts, imaginings, feelings and so forth. It seems clear that the value of the whole state of consciousness is not properly regarded as the sum of the value of its parts. Moore's view is that this is because, although each individual component of the total consciousness has its own value, one which it could retain in other contexts, the consciousness as a whole has a further value. But if a total consciousness is an organic whole, then some or all of these parts could not exist in the same character in another different sort of whole. In that case the reason the whole has a value other than the sum of the values of its parts, considered as capable of

** His only at all promising reason seems to be that if a thing owed something of its character to being a part of a whole, then we could not identify it independently and say of it that it is a part of the whole. In other words, it has to be something in its own right in order to join with other things to make up a whole. This argument seems to rest on the false assumption that one cannot identify something without knowing its complete character and all that is implied by that complete character. But why should one not know enough about something to identify it previously to saying that it is a part of something, without knowing all that follows from its being such a part?

existing unaltered in other wholes, may be that it is a radical mistake to think of them as capable of so doing.

Some have thought that the life of a society can have a value greater than the sum of the values of the lives of the individuals composing it. Moore, who did not place much value upon a society as a whole, might not have agreed, but his general principles allow it as a possibility. But may this not be because a society is an organic whole and the members could not live these lives except in it? If so, we can allow that in some sense the value of the society is a matter of the values actualised in the distinct lives, while insisting that these values could only be actualised within just such a society. This seems a more satisfactory way of looking at the possibility that a society may be an organic unity than Moore's principles allow.

The principle of organic unities is wielded as yet another weapon against hedonism. Moore grants that all very great goods are organic unities which have pleasure as a part. It is true of these wholes that if they remained otherwise the same, except that the pleasure was removed, they would only have at most a slight value. The hedonist's mistake is to assume that if X + Y has 100 degrees of goodness and X without Y has 10 degrees of goodness that Y possesses 90 degrees of goodness. But if X + Y is an organic unity then the fact that it has 100 degrees of value may be because it has 90 degrees of value as a whole while Y has none. So Y may have no worth in itself and yet *contribute* 90 degrees of value to X + Y, as being necessary to its existence as a whole. It is not a mere means to X + Y, since it is a part of it, but still the 90 degrees of value are not *its* value. Moore claims that this is precisely the role played by pleasure in all very great goods with which we are familiar.

Moore even allows that it is a plausible, though mistaken, supposition that all good things include pleasure. Thinking this to be the case one might infer that pleasure was the only good thing. For if one supposes that all good things become valueless if you remove the pleasure, one may infer that all the value must lie in the pleasure. But this would be a faulty inference since the following is quite possible: X has no goodness, Y has little or no goodness, but X + Y has very great goodness.

7. Moore's approach only makes sense in the context of his belief

that any proper thing can be conceived independently of any larger context in which it may figure. This becomes still more apparent when one turns to his other main methodological principle, *the method of isolation*. According to this, we can best assess the intrinsic value of something if we ask how good or bad a thing it would be if it existed in complete isolation. By doing this, we separate off everything which counts for or against it in virtue of its effects, or alternatively, its contribution to the formation of certain wholes.

This does seem very reasonable initially. What could be more sensible than the suggestion that the best way of evaluating a thing for its very own self, and not for its effects or its contribution to the value of larger wholes, is to consider what value it would possess if it existed in complete isolation?

Consider how this method might be used to test the value of *an Arts education*. Those who argue about this (about how much should be spent on Arts faculties in universities, for example) tend to debate whether it does or does not help the economy, help people to do various jobs, and so forth.

The spirit of Moore's approach to such matters is salutary. It implies that we should decide first whether we are considering its value as a means or its value in itself. The usual sort of discussion seems to concern its value as a means. But we cannot deal with this without asking ourselves what the results of a university education are supposed to be and what degree of intrinsic value pertains to them. The answer may turn out to be that the main results of university education for which intrinsic value can reasonably be claimed – such as the activity of critical thought – are included as main elements in the educational process itself, so that it is pointless to go on putting essential questions off by starting with questions about the value of the results of an Arts education. It will be better to leave that question, together perhaps with questions about the value of larger totalities of activity of which such university education is a part, until after we have considered what value it has in its own right. This will alert us to the error in assuming that the sole way of justifying spending money on courses in the Arts must lie in claims about their utility for ends beyond themselves.

But how can we decide how far an Arts education is good in itself? Moore's method of isolation bids us try to conceive a sample of such education in entire isolation. How on earth are we to do this? Perhaps we might try to think of an ideal university tutorial in an Arts subject as though it were a little world in isolation from everything else. We may then ask whether such a tutorial, perhaps taking place between angels with no past or future, summoned into existence for just this purpose, would be a *good thing*, and if so, how good?

This may seem a proper approach to a proper question,and yet it has an air of the ludicrous. Does it really make sense to conceive of a tutorial existing in isolation? Does the state of affairs include the existence of the tutor's big toe or not? Presumably it could exist without that, but if you pare away everything which there is equal reason to pare away, you are left with something very strange.

The truth is, surely, that we can form no reasonable conception of anything called education existing on its own with the kind of value we may like to ascribe to education in the real world. So Moore's method can hardly allow us to attach any intrinsic value to education at all; its value must be that of a means to other things or perhaps as an element in some larger whole of value. Yet education is very much the sort of thing you might expect Moore's principles to display as worth while in its own right rather than merely as a means to, or even component of, other things.

However, among the things which we tend to think of as good there are some few things which are more easily conceived in abstraction from any larger social or natural context and we will expect Moore's method of isolation to reveal these as the main bearers of intrinsic goodness. In contrast, the more experiences and activities are such that we find it hard to make sense of them existing or occurring in isolation, the more difficult it will be, on Moore's principles, to find anything of real inherent value in them. These expectations are fulfilled when we find that the chief things for which Moore claims great intrinsic value are personal friendship and the appreciation of beautiful objects.

An alternative very different conclusion which some might draw from the method of isolation is that only some vast whole like that of planet earth and all life on it at large can have intrinsic value. For it might be argued that nothing smaller can be imagined existing in isolation. We might even decide that only the universe as a whole has value. But Moore assumes we can ascribe inherent value to smaller units of reality and indeed this may be a necessary presupposition for the kind of calculations on the basis of which he thinks we should decide what is most worth doing.

That these implications of Moore's methodology seem strange might only show that it can guide us to unexpected ethical truths. However, if one feels that they are too strange one will wish to look carefully at some of the assumptions behind it, such as the denial of organic wholes.

8. Let us now consider more closely the main positive conclusion Moore reaches by application of his two principles, namely that personal affection and the contemplation of beautiful objects are by far the two most important good things of which we know. He contends that these seem to be the only things which are really very good when imagined in isolation. And we may well agree that that there is little else that one can imagine existing alone and being of any great value. We may agree also that their value is no mere summation of the values of their component parts, whatever precisely we take their parts to be.

Consider the variety of ways in which humans spend their time and ask which of them would have value if they occurred, miraculously, in a kind of vacuum. Most work in which humans engage with some personal satisfaction would seem emptied of all point thereby. The activities of politicians, of journalists, of doctors, of nurses, of postmen, of miners, of builders, of fishermen, of farmers, of salesmen, of advertisers, all seem to have their point and justification in results which lie beyond them. It is difficult to form any clear conception of what these activities would be like if undertaken in isolation but so far as one can form such a conception it is of something essentially futile.

What, for example, would the activity of a postman doing his work in a vacuum be? To imagine an activity in isolation presumably means imagining, so to speak, the minimum bit of reality which would have to exist for the activity to go on. So perhaps to imagine the postman's work in isolation is to imagine him walking down a street posting letters, or at least sealed envelopes, but imagining the houses as mere facades, with no rooms or people behind. If the postman worked in such a situation he would be insane.

Perhaps, however, we must think of the postman as having a seemingly well founded belief that he is delivering letters to real people. For what we should be doing is taking a bit of the real world and imagining it existing while the rest is blocked off, and the activities of a postman in the real world involve this sort of belief. To get the activity in isolation you have to think of someone suddenly coming into existence with just enough of a world around him so that what he does and believes is just what the postman does and believes in the real world, but without any real environment. But surely the activity of a postman in such a void would be utterly futile and valueless.

It may be said that this is not imagining the activity of a postman at all. For such activity can, as a matter of logic, only occur in a real country with a real postal system allowing epistolary communication between businesses, friends, relatives, parent and child, husband and wife, MP and constituent, etc, etc. But in that case the very idea of the postman's work occurring in isolation is incoherent since, as a matter of logic, it cannot be separated off from a whole range of activities beyond itself.

So either one cannot imagine postal activity in isolation, because it is conceptually bound up with so much else, or one can imagine it as the futile activity of a deluded loner. In neither case, does the activity of a postman emerge as something which could be good if it existed in isolation.

Try now the activity of nursing. You have to think of a nurse tending a sick person in a bed, though there is no real world outside the hospital ward. I am supposing we can include the patient as a real conscious person as 'part' of the nursing activity. If one cannot, one gets a deluded loner nursing a dummy. But even as it is, you get something rather odd. The patient has no destiny beyond whatever duration is required to get an episode of nursing, and she has no real past, nor does the nurse. Now such a little human episode, if the patient is not in serious pain, and the nurse enjoys her work, could have some value as a little stretch of human relationship, but all that really pertained to it as nursing would be futile, since the point of nursing is normally to return the patient to a healthy life in the world outside. (Even nursing of the terminally sick relates to the significance of death as the end of a complete human life, not just to the final days in hospital.)

Thus nursing and postal delivery both seem things which would lose all, or almost all, their value in isolation. They must, therefore, be dismissed as not being good in themselves, or not to any considerable degree. Their point lies outside them.

But is there anything which can be imagined as existing in isolation and as possessing great value in such a case? Well, if anything can, it seems to be Moore's two great goods.

One can imagine someone in a room looking at a beautiful painting, drinking in the details and the way in which they combine to constitute the lovely whole, and there being nothing beyond this observer and the object of his contemplation. Or one can imagine an experience of beautiful music existing in a void. Moreover, to imagine these does seem to be to imagine something with quite strong positive value.

The same goes for personal affection. One can imagine two people enjoying themselves affectionately together, and nothing else existing except the barest background to physical existence. At least one can make more sense of this than in our first two cases. By avoiding any practical direction to what they do together, one avoids the futility of the nursing situation where nothing exists beyond the hospital.

Are there no other types of thing for which Moore might have claimed a similar high value in isolation? What of mathematical thinking? What of artistic creation as opposed to contemplation? What of a game of chess or a game of cricket (without spectators)? However, even if he had added these, the list was bound to be short. Many leisure activities would hardly figure because we can only value them as ways of relaxing from engagement in what really matters. To satisfy Moore's requirements an activity has somehow on the one hand really to somehow matter, and on the other hand not to do so on account of anything ordinarily thought of as of practical use. Thus Moore's methodology points inevitably to his main ethical conclusion, which is that nothing, or at least very little, is to any great degree good except for cases of personal affection and the enjoyment of beautiful objects and that everything in life which does not come under these heads has barely any value apart from whatever it may have as a means of promoting these great goods.

Thus the point of the lives of those who are mainly engaged in politics, nursing, mining and so forth is to keep a society going in which personal love and the enjoyment of beauty can flourish. So far as their work goes it has point only as a means to these. Lovers and connoisseurs, on the other hand, are the ultimate point of civilization.

It is reported that an Oxford don, was asked, in the 1914 war, why he did not volunteer for the defence of civilization. 'I am the civilization for which they are fighting,' he replied. In the same way lovers and connoisseurs of beauty are what everyone else is, or ought to be working for, on Moore's scheme.

I suggest that both the strength and the weakness of Moore's positive ethical views stem mainly from his method of isolation.

For, first, this method springs from Moore's awareness of an important truth, namely that if we see the point of everything in its being a means to something else we empty the world of value. The thinking of politicians for whom education is only important if it helps boost the national economy, and this is important because it helps people enjoy what they want, and this is important because it encourages consumption and thus industry, either goes round in a vicious circle or takes off on an interminable regress. If anything is worth while in life at all, some things must be good in and by themselves.

But second, it nonetheless distorts our conception of human life. Seeing value in activities only in so far as we can conceive them retaining it when cut off from the main tides of human affairs, leads to a kind of preciosity and detachment from what excites most human beings which is ultimately impoverishing. How ethical theory might do justice to both these points remains to be seen.

9. The main moral upshot of Moore's ethics is a certain view as to what is really worth while in human life. Yet Moore did not think value could only occur in relation to human experience. This was not because he had any interest in values realized in animal life, but because he believed that some degree of goodness pertained to things or states of affairs which do not involve consciousness of any kind. There is a famous passage in which he invites us to

acknowledge that it would be better that a world of which no one would ever be conscious should contain beautiful objects than that it should contain ugly objects.

The appeal of this lies, I suggest, in the fact that if we imagine a beautiful object, we cannot help thinking it good that there should be such a thing. The dubious part of the claim lies in Moore's belief that beautiful objects with their beauty and the qualities which make them beautiful, could exist without any consciousness of them. To me it seems that beauty, and indeed the qualities and forms which give rise to beauty, only exist for a consciousness. The objection to Moore is not so much that beauty, without consciousness of it, would not be good as that beauty, without consciousness of it, would be impossible. However, Moore's claim deserves to be taken seriously by those who think that the objects we perceive exist just as we perceive them when unobserved. Yet I am puzzled as to whether that would mean that we should leave our radios on, so long as the music coming from them is beautiful, in order that our rooms be filled with beauty even in our absence.

Although he holds that value can be realized in the absence of consciousness, Moore does not believe that situations not involving consciousness can be *very* good. Or rather, he holds that all the great goods of which we know involve consciousness; there may conceivably be others of which we are ignorant. Certainly personal affection and the admiring contemplation of beautiful objects involve consciousness, although it would be a mistake to think that there is no more to their value than comes simply from the kind of consciousness involved.

The enjoyment of beauty is, in a sense, the more basic of Moore's two great goods. For personal affection, insofar as it is good, consists mainly in the admiring contemplation of good states of another person, and these must either themselves be instances of personal affection, requiring good states of someone else for their object, or be instances of the love of beauty. Sooner or later therefore the goodness of personal affection depends upon the existence of someone's appreciation of some object's beauty.

10. What becomes of the rightness and wrongness of actions, and of duty, on this scheme? These, for Moore, have little or no

intrinsic value, and their investigation belongs to the practical branch of ethics concerned with which among possible actions will have the best consequences, rather than to the more fundamental enquiry into the nature of the intrinsically better and worse which it presupposes. For, according to Moore, to say that an action would be right is to say that the total results of doing it would be at least as good as would be those of doing any alternative. To say that it is a duty to do it is to say that its consequences would be better than those which would ensue from any alternative. When there is more than one right act it is one's duty to do one of them, though not one in particular.

Thus Moore's position on the rightness and wrongness of actions is a form of rigoristic utilitarianism in which effects in terms of intrinsic good and bad replace effects in terms of pleasure and pain. It is often known as ideal or agathistic utilitarianism.

Moore gives some interesting twists to usual utilitarian themes. He puts great emphasis on the difficulties of prediction, and urges that where there are rules to which people do in fact adhere for the most part, and which help maintain the social stability required for any kind of good to flourish, we are likely to come nearest to doing what is objectively right (in terms of its actual consequences) if we also stick to the rules, but that where the rules, however useful they would be if generally obeyed, are widely flouted we should make a direct judgement of what will have the best consequences. Where there are no established rules to guide decision we must make a direct judgement as to what will have the best consequences, concentrating on the more immediate and those we are personally most motivated to promote, as the remote future is highly unpredictable and we are unlikely to do much to promote what is not personally inviting. These however are directives as to what is likely to be the right action; the actually right action is that which will actually, not merely probably, have the best consequences.

We may note that although Moore thinks that in some sense what one *should* do is what looks most likely to be one's duty in the sense of that which will produce the best consequences, it is unclear what meaning he can ascribe to this 'should'. He cannot equate it with 'ought' since, according to him, to speak of what one ought to do is simply to speak of what *will* have the best consequences. Some philosophers have distinguished between our 'subjective' and 'objective' duty, and said that our subjective duty is to do what we think most likely to be our objective duty. ('Subjective' here has nothing to do with ethical subjectivism; it is generally supposed that the existence of this subjective obligation is an objective fact.) However, even with this distinction, there is some difficulty in capturing the imperatival force of the 'should' in an ethical system in which intrinsic value is the only basic ethical concept.

What is strikingly different in Moore's treatment of practical ethics from that of the classical utilitarians is the lack of any reforming zeal. This goes with the tendency to concentrate discussion on the intrinsically good, rather than the intrinsically bad. He does, indeed, say something about the latter. Pain, or at least the consciousness of pain (for Moore professes to distinguish these), is said to be very bad, but the main other things which he regards as bad are the enjoyment of ugly things, and the dislike of beautiful things! There is no suggestion that the primary ethical task is to relieve the immense weight of human suffering.

11. For quite some time after publication of *Principia Ethica*, its type of ethical intuitionism seems to have been the dominating moral philosophy in Britain, though not in America. Main intuitionists included C.D. Broad, W.D. Ross, A.C. Ewing. Most ethical intuitionists came to think that what the doctrine of the naturalistic fallacy established was not so much that *good* is indefinable as that ethical expressions are either indefinable or only satisfactorily definable in terms which involve some other distinctively ethical expression. They held that there is a limited set of fundamental ethical terms, such as *good*, *bad*, *right*, *wrong*, *ought*, *duty*, such that you cannot explain what one of them amounts to except either by appealing to intuition of some simple non-natural property, or relation, to which it refers or by defining it by way of some other ethical term of which this is true. The main debate was as to which of these ethical expressions were most basic. One view was that they form a kind of virtuous circle of equally basic expressions definable in terms of each other. This idea has its problems. One might well think that a word cannot both label a property or a relation revealed to intuition and its meaning be a construct out of the meanings of other expressions.

However, it seems possible that several of these words should label distinct properties or relations which however are necessarily related to each other in ways which the definitions articulate. (Similarly geometrical shapes or relations may be interdefinable in terms of their necessary relations and yet each, in principle, be displayable to 'spatial intuition'.)

W.D. Ross, for one, fully endorsed the doctrine of the naturalistic fallacy, but he held that it was a bad mistake on Moore's part to define such expressions as 'right', 'wrong', 'ought' 'obligatory' by way of 'good'. For Moore, to say that an act was right was to say that no alternative action would produce better consequences, while to say that it was 'a duty' or 'obligatory' was to say that it would produce the best consequences possible. Ross objects that Moore himself was guilty of confusing a contestable moral judgement in favour of his ideal utilitarian position with an account of the meaning of an ethical expression in much the same way as were those who committed the naturalistic fallacy by offering naturalistic or metaphysical definitions of 'good'. Just as there is a substantial move from saying 'This is a pleasant experience' to saying 'This is a good thing', so there is a substantial move from saying 'This act will produce the best consequences' to saying 'I ought to do this act'.

To see the reasonableness of Ross's claim one need only consider a case of which Ross made much, that of the duty of promise keeping. Surely someone can meaningfully say 'I ought to do this, because I promised, even though better consequences would ensue from my doing something else which would involve breaking the promise.' An ideal utilitarian like Moore may claim to have the moral insight that promise keeping is only right or obligatory, where one cannot do better or as well by breaking the promise (taking general account of effects on human trust into account) but he cannot claim that this insight is merely into how words are properly used. Moreover, just as Moore thought that freedom from the naturalistic fallacy made one more ready to recognize that there is a plurality of basically different sorts of good thing, so Ross thought that freedom from this wrong definition of right and duty in terms of good made one ready to recognize that there is a plurality of grounds of obligation, of which the obligation to produce good and reduce bad is just one, or rather two, for he thinks the duty to prevent bad a distinct, and usually more stringent, duty than to produce good.

For Ross obligatoriness or rightness, as a property of possible actions, is the most basic ethical property or relation. He held that there were a variety of different rightness-making properties. We recognize by moral intuition that certain 'natural' properties of an action will make it obligatory. The non-natural characteristic of obligatoriness is 'supervenient' upon each of these natural characteristics. It arises out of them but is quite distinct. Right-making or obligatoriness-making characteristics include that of being the keeping of a promise, being a case of truth telling, of gratitude, or reducing bad or evil, of promoting good, of promoting knowledge.

Thus Ross echoes earlier rationalist moralists who held that the basic general principles of duty are self-evident and necessary truths revealed to rational intuition. A standard objection to such views is that these general principles often clash one with one another as necessary truths could not. Sidgwick, for example, argued thus that the typical duties recognized by common sense could not be self evident and necessary since the clashes between them needed to be resolved by appeal to a more fundamental principle, that of the principle of utility, and that it was therefore this alone which possessed such a status.

To meet such objections Ross developed a very useful concept which has become part of the regular stock in trade of moral philosophers, the concept of a prima facie duty. One can say (he held) that it is absolutely true that a certain characteristic always gives rise to the property of prima facie obligatoriness. An act which has this characteristic is a prima facie duty. However, if there are one or more alternative acts which also possess some characteristic which similarly gives rise to prima facie obligatoriness, then only one of them can be all-things-considered obligatory.* In order to be all-things-considered obligatory a prima facie obligatory act must not be incompatible with some more stringent prima facie obligatory act. As to which is the more stringent obligation in a situation in which more than one

* The actual phrase all-things-considered obligatoriness is not used by Ross. I have borrowed it from *The Nature of Morality* by Gilbert Harman.

applies, that is for intuitive assessment in the light of all that can be known about the particular circumstances of the case. Thus intuition comes in twice, firstly to tell us general truths to the effect that certain natural characteristics bestow prima facie obligatoriness, and secondly to tell us which of various alternative acts is more stringently obligatory. In both cases, however, we are intuiting the necessary determination of ethical characteristics by 'natural' characteristics. Even when we are balancing one consideration against another in a quite particular case, to decide which of two conflicting prima facie obligations is the more stringent, we are trying to reach such an intuition, but the wealth of detail which determines the relative degrees of stringency in this case will make it inexpressible in a formula mechanically applicable to other cases. A superficially similar case may differ from this one in some subtle matter of detail the relevance of which we will only recognize when we encounter it in the concrete.

The prima facie duties Ross recognizes include the duty to maximise good, and the usually more stringent duty to minimise bad or evil. Good and evil are understood, just as with Moore, as non-natural characteristics whose presence in things is revealed to intuition. However, the duties to promote good and minimise evil may sometimes be outweighed by other duties, such as that of keeping a promise or a duty of gratitude, which in the particular circumstances are more stringent.

It does seem that Ross's view is (as he claimed) very much that of the man in the street (or at least in the streets with which Ross was familiar). Indeed, Ross thinks the moral philosopher must take the deliverances of common morality as his basic data (as Moore, in spite of his much proclaimed faith in 'common sense' did not) though this is because it rests on intuitions of necessary truth. True, the moralities of many societies are defective, but this springs either from defective factual information or from the limited extent to which people have begun to engage with certain issues in a truly moral way. As the human mind matures the same basic insights will impose themselves on all.

Ross's view that duty is not totally determined by consequences is usually called the deontic position, and is contrasted with consequentialist positions of which hedonistic and ideal utilitarianism are the chief examples. Ross is probably right that his is the more common sense view, but it may still strike us as extremely odd that it could ever be one's duty not to do that which would certainly have the best results.

The answers offered by intuitionist philosophers such as Moore and Ross to the three questions about ethical propositions listed in the introduction are clear. Ethical propositions are objectively true or false. The true ones tell us what possesses, or leads to what possesses, certain uniquely ethical properties. There is a rational means of deciding whether to accept them or not. For Moore this is, in the case of propositions about what is intrinsically good or bad, a matter of applying the method of isolation, guided by respect for the principle of organic unities, and discovering what suggests itself to intuition, while for judgements about right and wrong in action it is a combination of empirical assessment of consequences of action (using induction or whatever else is considered suitable here) combined with judgements of intrinsic good and bad. For Ross the most important method is that of testing ethical principles by reflective intuition. If, like Moore and Ross, you think that intuition, in their sense, is a form of rational insight, there is no reason not to regard this as a rational method. Finally, so far as the question of convergence goes, Moore insists that there is no guarantee of it. If our intuitions differ, one of us at least is wrong, but we may be unable to find out which. However, although he says this, partly to emphasise that the truth of the matter is quite independent of the question whether we know it, it would seem reasonable to expect that, if there is a real truth here into which we might have rational insight, careful attention to just what is at issue will bring about convergence. This, indeed, seems to be what one would expect from the point of view of intuitionist philosophers in general.

CHAPTER III

The Attitude Theory of Ethics and Some Related Theories

1. Moore's claim that definitions of ethical expressions in naturalistic terms commit the 'naturalistic fallacy' long continued to be a main influence on moral philosophy in the English speaking world. But quite soon different conclusions were drawn from this than Moore's. Moore and other intuitionists thought that, since the most basic sorts of ethical statement did not say anything true or false about matters of natural or empirical fact,* they must say something true or false about some special ethical non-natural fact. By the 1940's it was becoming more usual to conclude rather that their distinctive role was quite other than that of saying anything factual at all. This thesis is sometimes called non-cognitivism. This expression covers emotivism, of which C.L. Stevenson and A.J. Ayer are the best known exponents, and the prescriptivism of R.M. Hare. The latter received some notice in chapter one (and will be discussed briefly again in chapter five); in this chapter I shall be concerned primarily with the position of Stevenson, whose *Ethics and Language* is the fullest statement of the doctrine.

Although Stevenson's main work, *Ethics and Language*, is supposed to be well known, it is an under-rated and often misrepresented work, accounts of which in books on moral philosophy are often little more than parodies. However, in my

* It should be borne in mind that Moore also rejected the idea that ethical statements were especially concerned to state some metaphysical matter of fact. The non-natural ethical facts in which he believed were not concerned with the existence or character of metaphysical entities such as pure egos, the Absolute, God or the soul, but with the distinctively ethical form of the good. Other intuitionists and the non-cognitivists equally opposed metaphysical interpretations of ethical claims. However, I shall leave this to be understood for the most part.

opinion, the best parts of it are those which present not so much the strictly emotivist thesis as a more general attitudinist thesis which is more convincing when detached from the former.

Emotivism claims that the distinctive meaning of ethical words lies in their power (in virtue of our early linguistic conditioning) to stimulate emotions and attitudes in a hearer, and vent them on the part of the speaker. This thesis about the meaning of ethical words is presented against the background of a *causal* theory of meaning in general. According to this, the meaning of words and statements lies in the psychological states (including long standing dispositions) which they combine to produce in hearers and to be caused by in speakers. Words have *descriptive meaning* insofar as their causal connection is with the psychological states known as beliefs, and *emotive meaning* insofar as their causal connection is primarily with attitudes. 'It is raining' is typically caused by and causes the belief that it is raining, and therefore has primarily descriptive meaning. In contrast to this, according to the emotivist thesis, the typical cause and effect of a statement like 'Personal affection is a great good' is not any kind of genuine belief, which could be true or false, but an emotional attitude of favouring personal affection, which each of us may find ourselves either sharing or otherwise, but which we cannot properly call true or false; it therefore has primarily an emotive rather than a descriptive meaning. Likewise the word 'good' serves primarily to determine the sorts of attitude sentences are causally associated with, while the word 'cat' serves primarily to determine the sorts of belief with which statements are causally associated. They thus contrast as words one of which has primarily emotive, the other of which has primarily descriptive, meaning.

There are various grounds for dissatisfaction with these notions. It seems doubtful, for one thing, whether we are conditioned so that words have this kind of direct causal power over us. Surely if someone says 'The earth is flat' that does not even begin to cause a belief to that effect in me, nor does someone saying 'Cruelty is a good thing' even begin to cause approval of cruelty in me. It is much more satisfactory to say that these statements show what the speaker believes or feels, if he is speaking sincerely and correctly, and invite the hearer to an exchange of views in which he will agree with him or otherwise in belief or attitude. Where a statement does cause the relevant belief or attitude in a hearer, it is because he has some reason to think the speaker's beliefs on that topic are likely to be true, or his attitudes ones he is likely to find, on enquiry, reason to share, rather than because the mere words have any great hold over him.

Another reason for rejecting this causal theory of meaning is that when one asks what is meant by a statement one is normally concerned to know what a reasonable speaker is saying in uttering it. What the actual or even likely or standard effects of his saying this would be on a hearer, himself included, has little to do with what a reasonable speaker means in saying something. For the speaker cannot settle what he means by settling on the effects it will produce as he can settle what he is setting out to mean in saying it.

Suppose one asks whether a speaker is using 'good' in an ethical sense or not. The answer cannot be given by considering whether in the context of utterance it is likely to produce attitudes of some distinctively ethical sort, rather than some non-ethical attitudes or perhaps beliefs. Nor can it be a matter of the effects he intends to cause in hearers, for one may say something with a definite meaning, with all sorts of different such intentions, and, in any case, the causal theory as advanced by Stevenson is concerned with the actual causal potency of language, not with what is intended.

It may seem more plausible to identify the meaning with the belief or attitude of the speaker which caused it. However, one cannot satisfactorily identify the question what a speaker meant in saying something with the question what caused his utterance. If the speaker is lying, for instance, it may be that what caused his utterance was something quite opposed to belief in what he meant to say, or a favourable attitude towards what his utterance was meant to commend. In general, the actual psychological causation of an utterance may be related to all sorts of different degrees of endorsement in one's actual beliefs and attitudes of what one is meaning to say. True, we should distinguish between what a particular speaker means and what his statement means. If I use a word incorrectly what I mean will not be what my statement means. However, the two must not be contrasted too sharply. The meaning of a statement seems to be roughly what the speaker should have meant by it and this must in most cases be what the speaker does mean by it. But what he does mean by it cannot be identified, at least in any simple way, with the actual beliefs or attitudes which caused it.

Doubtless words do sometimes have emotive effects, and some ethical words may have fairly standard such effects in certain communities. Equally, emotional attitudes may figure as standard causes of certain sorts of utterance. The tendency to have such causes and effects can be called emotive meaning, if one likes. But it is a mistake, surely, to think of emotive meaning as a form of meaning in that most important sense of meaning in which we ask what someone meant by saying something, or what he would have meant if he had been speaking properly.

In order to have what seems the best of Stevenson's thought before us, I shall now develop an attitude theory of ethics, largely based on his work (though also drawing on that of A.J. Ayer), in which the strictly emotivist thesis is dropped and the concept of emotive meaning gives way to one of valuational meaning. I shall ascribe it to an imaginary person whom I shall call the attitudinist.

2. The attitudinist begins with some reflections on the meaning of straight factual statements. (By a statement he means roughly a declarative sentence as used, or as it might be used, on some particular occasion, so that indicator words like 'I' 'this' and 'now' are given a definite reference.) He says that they express beliefs. Thus the statement 'It is raining' expresses the speaker's belief that it is raining and the statement 'Bertrand Russell was a writer of short stories' expresses the speaker's belief that Bertrand Russell was a writer of short stories. In virtue of expressing the belief that it is raining the statement 'It is raining' says or asserts that it is raining. One might say that it claims that it's presently raining is a real state of affairs.

One could also say more impersonally that the statement 'It is raining' expresses the belief, not merely the speaker's belief, that it is raining. That makes it clearer that someone else who assents to (or dissents from) the statement is expressing their belief (or disbelief) that it is raining. If one's companion says (a little artificially) 'Quite so' (perhaps 'Quite so, what of it?') she is expressing her belief that it is raining. In short, related to a factual statement is a certain belief of which both those who utter it and those who assent to it express or exhibit their acceptance.

To say that a factual statement expresses a belief is not necessarily to imply that the speaker, or the hearer who assents to it, really has that belief. They may only be pretending. It is simply a rule of the language, or rather it follows from its basic rules, that one speaks misleadingly if one utters or assents to that statement without having that belief.

When a statement expresses a belief, it does not say that the speaker (or one who assents to it) has that belief. To assert that the knave of hearts was the person who stole the tarts is not to assert that one believes that it was he who stole them. When someone says 'It was the knave of hearts who stole the tarts' I agree with him or not according as to whether I believe that it was the knave of hearts who stole them, not according as to whether I believe that he believes that it was the knave. One could say that the sentence 'It was the knave of hearts who stole the tarts' puts the state of affairs of its having been the knave of hearts who stole the tarts upon the mat for discussion. We are invited to consider whether it is a real state of affairs. It does not put upon the mat (not at least in the same direct way) the state of affairs of anyone believing that it was the knave who stole them.

If 'It is raining' expresses the belief that it is raining, but says not that I, the speaker, have that belief, but simply that it is raining, it would appear that it differs in meaning from 'I believe that it is raining'. Moreover, it would appear that the difference is that the latter sentence expresses my belief that I have the belief that it is raining, and says that I have the belief that it is raining, and does not express the belief that it is raining or say that it is raining. This would be so if 'I believe that it is raining' were a psychological statement about myself. (Perhaps one should say that as such a psychological statement it would express my *awareness that I have the belief* rather than *my belief that I have the belief*, since belief is not quite the right word in this context, but let us leave such qualifications to be understood.) If the statement is taken in this way, a hearer should assent to it only if he believes that I have the belief. He need not concern himself with whether it is actually raining.

Statements opening 'I believe that . . .' are probably on occasion used in this way. One might be asked to report upon one's beliefs to a psychologist or sociologist, interested in current opinion on matters biological, and say 'I believe that natural selection does not adequately account for the evolution of animal life'. One would then not be saying that natural selection does not adequately account for the evolution of animal life but that one has a certain belief to the effect that it does not do so. Of course, if one has the belief one would indeed be prepared to say that it does not adequately account for the evolution of animal life, but that is not what one is saying here and now.

However, although it is possible to use statements opening 'I believe that . . .' to make a psychological report, that is not their only or their most usual use. For usually they say, rather tentatively, that things are as specified in the wording which follows. Thus if I say 'I believe it is raining' I am usually saying, rather tentatively, that it is raining. What I am expressing is a tentative belief that it is raining, not the belief (or awareness of the fact) that I believe that it is raining.

Perhaps it oversimplifies the situation to treat these as two quite different uses of such expressions as 'I believe that . . .' It seems, rather, that it is built into the meaning of 'I believe that . . . ' that it hovers between expressing tentative belief that what is specified by the following wording is so, and expressing belief or awareness that the speaker believes that it is so. For our purposes, however, no harm will be done if we distinguish two uses of 'I believe that . . .', one in which it expresses the tentative belief that what is specified by the following wording is so, the other in which it expresses the belief or awareness that the speaker has the belief. The first could be called the modal use, because it has the same kind of force as the modal word 'probably', and the second the psychological use because it makes a statement about one's own psychological state of belief. This distinction will have some importance when we consider the grounds on which the attitudinist distinguishes his position from that of ethical subjectivism.

The attitudinist now asks us to consider whether there may not be sentences which express, in the sense of 'express' in which factual statements express beliefs, something other than beliefs. Two initially promising examples are interrogative and imperative sentences. Perhaps the sentence 'Is it raining?' expresses the state of wondering whether it is raining, and the sentence 'Shut the window, please' expresses the state of wishing the hearer to shut the window.

It sounds promising to say that these sentences stand to wonderings and wishes very much as factual statements stand to beliefs. However, it seems that the relation is not quite the same, for there is no comparable way in which a hearer can express the same wonder or wish by assenting to the sentence as uttered by the speaker. 'Shut the window, please' is said in a situation where the speaker rather expects the hearer to act so as to fulfil a certain sort of wish of his, if he indicates that he has it by an imperative sentence. It would be inept, rude, insubordinate, or whatever, simply to say 'I agree' or nod to show that one also wished that the window was shut. There are somewhat similar ways in which it would be odd to express assent to a question to show that one shared the speaker's wondering. So if imperatives and interrogatives express wishes and wonderings it is in a somewhat different sense.

3. It seems, then, that sentences which are not declarative cannot express anything in the sense in which factual statements do. That leaves open the possibility, however, that some declarative sentences or statements are not factual and express something other than beliefs. The proposal of the attitude theory of ethics is that there is another class of declarative statements which express so-called attitudes, in just the sense in which straight declarative factual statements express beliefs. By an attitude is meant any way of being in favour of or of disfavouring (of being against) something. The most significant attitudes are those which either involve, or consist in, the kind of serious wishes that certain things should be done or left undone, which, in appropriate circumstances, will produce actual actions directed to that end.

The attitudinist calls statements which express attitudes rather than beliefs value statements and he thinks of ethical statements as a more or less vaguely delimited subclass of these. Actual attitude theorists have on the whole taken rather lightly the task of distinguishing the particular kinds of attitudes expressed by those value statements which are distinctively ethical. One suggestion is this. The attitudes expressed by value statements are all ways of favouring or disfavouring things which we would like to see shared. They contrast with such attitudes as we are content to leave as our own merely personal inclinations and which are therefore not given the dignity of expression in a statement. However, not all attitudes expressed in value statements have the requisite stridency to be called moral. These more strident attitudes expressed in properly speaking ethical statements are ways of being in favour of or against types of behaviour with a degree of force which makes us wish disfavoured actions discouraged by some kind of social sanction. This social sanction will be, in the first place, some kind of social ostracism or discomfort; it may or may not be wished that it be associated with a legal penalty.

Another possibility is that we call statements expressing attitudes with this particular sort of stridency *moral* statements and allow as *ethical* all statements which express attitudes towards conduct of a certain special seriousness and pervasiveness in their influence on one's own behaviour and such as one would like to find widely shared, but not necessarily to have supported by a social sanction.

However that may be, we must now note that value statements in general, and ethical statements in particular, are distinguished by the attitudinist into two types. There are those which, like 'Stealing is wrong', express a pure attitude, without any admixture of belief. Of course, my attitude may stem from my having certain beliefs, but these beliefs are not part of what is actually expressed by the sentence, and do not pertain to that which someone else must share with me, in order for it to be logically proper for him to assent to my sentence. There are others which express an attitude towards some thing or state of affairs the existence of which is stated, or at least implied, and which therefore also express a belief in its existence. 'It was wrong of you to steal that money' is of this latter kind. However, even though it expresses my belief that you did steal the money, the distinctively ethical role of the sentence is to express my attitude of disfavour towards you for this act.

The great contrast, according to this moral philosophy, between beliefs and attitudes is that beliefs can be true or false, while attitudes cannot be. If someone believes that it is raining, or that the knave stole the tarts, there is either a state of affairs such as he believes in or there is not, and his belief, and the statement which expresses it, is true or false accordingly. When a sentence expresses an attitude or a wish it cannot be true or false in this sense, for wishes and attitudes cannot be true or false as beliefs can. Wishes can of course be fulfilled or remain unfulfilled and things may or may not be as people with certain attitudes would like, but it would be thoroughly misleading to speak of truth or falsehood here. Beliefs are acceptable or not according as to whether they succeed in their goal of conforming to reality and receive credit or discredit for this by being called true or false. The acceptability of wishes and attitudes is not based upon such conformity, though of course they normally lead to the effort to produce it.

The attitudinist will admit, indeed, that there is a weak sense of 'true' and 'false', in which they merely register agreement or disagreement, and in which it is quite legitimate to use them to express agreement or disagreement in attitude with an ethical statement. But we are mistaken if this leads us to suppose that there is any question of truth or falsehood here in the sense of conformity to fact.

4. It is important to realize that when the speaker uses an ethical statement to express an attitude, the hearer who agrees or dissents is not agreeing to, or dissenting from, any claim to the effect that the speaker has the attitude. Rather, he is expressing his own agreement or disagreement in attitude, and to agree or disagree in attitude is not to agree or disagree about an attitude. This is implied in saying that the original statement expressed an attitude, and not a belief about an attitude.

It is on the basis of the distinction between statements which express attitudes and those which state that one has an attitude that the attitudinist thinks his theory significantly different from ethical subjectivism as usually understood. For by this is usually meant the thesis that ethical statements are reports upon, statements about, one's own feelings or attitudes. If so, an ethical statement made by me would be true, and could not reasonably be challenged by others, if I have the feelings in question. Certainly they might have different feelings, and could report upon them in their own ethical statements, but they could not object to my ethical statements as I meant them, if I have the relevant feelings. But for attitudinism an ethical statement does not *state* any fact at all, not even the fact that the speaker has a certain attitude; rather it *expresses* an attitude, something, so it is contended, which is very different. Just as 'Prostitution is on the increase' is quite distinct in meaning from 'I believe that prostitution is on the increase', at least when this latter is taken as a psychological statement rather than as a modally qualified form of 'Prostitution is on the increase', so 'Prostitution is wrong' is quite different in meaning from 'I disapprove of prostitution' at least when this is taken as a psychological statement. It could only be assimilated to it if 'I disapprove of prostitution' were treated as a tentative way of saying that prostitution is wrong, that is, as a tentative expression of an attitude against prostitution rather than as an expression of one's awareness of one's own disapproval of it.

The distinction between subjectivism and attitudinism may seem a mere verbal subtlety. However, though it may not be as great as has sometimes been maintained there is a genuine difference. For one thing, the attitudinist view escapes an objection to subjectivism made by Moore that if when I say an action is bad I report my negative feelings towards it, and when you say it is good you report your positive feelings towards it, then there is no disagreement between us (since we will probably both agree that the other has the feelings he reports) which Moore thought an absurdity. Stevenson* commented on this that we disagree in attitude, but not (necessarily) in belief, and in any case, for attitudinism, since the two statements express opposite attitudes, they are incompatible in much the same way as that in which statements which express opposite beliefs are. More generally one can say that for attitudinism ethical statements invite the hearer (or perhaps the speaker when in discourse with himself) to express his agreement or disagreement in attitude, or perhaps join in a shared attempt to arrive at some stable and better based attitude, towards what is in question, and that this is quite a different matter from any discussion concerning the

* See *Facts and Values* Chapter VII. Stevenson's relation to subjectivism is not quite that of the attitudinist as described above (which here is closer to Ayer's position) since (at least at the time of *Ethics and Language*) he thought that ethical statements did often state that the speaker had certain feelings, though they went far beyond doing merely this in virtue of the emotive meaning of the ethical words.

psychological state, of approving or disapproving, on either the speaker's or hearer's part.

5. Although (according to attitudinism) ethical statements express attitudes rather than beliefs they are in many ways quite comparable to factual statements which express beliefs. They express a certain mental stance to the world which others are invited to accept or reject just as do factual statements, and as mere imperatives, exclamations and interrogatives cannot. Moreover (as we shall see shortly) they can enter into standard logical relations with factual statements.

However, there is a disanalogy in one respect. If I say 'Prostitution is on the increase' I express my belief that it is on the increase, but what I put upon the mat for discussion is not my holding the belief, but rather the state of affairs, which I hold to be actual, but which others may not, of prostitution being on the increase. What do I similarly put upon the mat for discussion when I say 'Prostitution is immoral'? A verbal answer, which will do as part of the language game of ordinary life, is that I put the immorality of prostitution upon the mat in a way quite comparable to putting its increase on the mat. However, for the present attitudinism there is no state of affairs, which may be real or unreal, of prostitution being immoral, as there is a state of affairs, which may be real or unreal, of its being on the increase.

It is sometimes supposed that if this is so, there can be no rationality in ethics. For is not rationality concerned with assessing evidence for claims to the effect that certain states of affairs exist? But if there are no such things as ethical states of affairs, then there can be no question of evidence for them. Stevenson himself goes along with this view to a far greater extent than seems appropriate. For, with some qualifications which do not much affect the main point, he suggests that when we give reasons for an ethical statement we are typically making factual claims acceptance of which we hope will cause others to have the attitude it expresses and he distinguishes this sharply from the case where one statement gives some kind of logical or rational support to another. (He does not even make any special point about the dishonesty which there would seem to be in doing this when acceptance of those claims is not the cause of one's own attitude – to do so would be, from his point of view, to play the moralist rather than the philosopher who should simply describe ethical discourse.)

Of course, if we make factual claims in support of an ethical statement, then these can be evaluated rationally, by way of induction and abduction. Thus so far as difficulty in settling for ourselves, or agreeing with others, on the right answer to an ethical question turns on disagreement or doubt about factual matters, agreement on which would cause settled or agreed ethical attitudes, there *is* a place for reason in ethics. However, wherever there is hesitancy or opposition in ethical attitude which is not rooted in hesitancy or disagreement in belief, in the sense that no amount of factual certainty or agreement would cause the final ascendancy of one definite attitude, Stevenson thinks that the ethical question cannot be resolved rationally. What remains is a possible non-rational resolution of the question, which may consist in one side to a dispute stimulating a change of attitude in the other by a moving use of emotive language, or one aspect of one's personality achieving dominance over another. Still, it is also quite possible that hesitancy or disagreement may persist without one side of the dispute or one aspect of one's personality being more rational than another, or in possession of any deeper insight into truth.

Yet there are indications of the approach to ethical questions which, from Stevenson's point of view, would be the most rational. It is essentially a matter of looking into all the factual issues which one believes likely to affect one's attitudes on the issue, or which are likely to affect the attitudes of those debating it with one. One then lets the influence of the beliefs on all these matters at which one arrives play upon oneself and/or others, until this causes settled, and with good fortune agreed, attitudes to emerge in all concerned. However, this good fortune cannot be guaranteed and human psychology may be such that we are doomed to permanent hesitancy or disagreement, a fact with which each will have to cope from the perspective of his own attitudes. This attempt to be influenced by the maximum of factual knowledge is in a sense a rational way of trying to answer ethical questions (this being perhaps the Stevensonian answer to the second question raised in the introduction), but it offers no guarantee of congruence. Thus our third question is answered negatively.

But if this is to be called a rational method, it is so, from Stevenson's point of view, mainly because it consists in letting one's attitudes be moulded by rational factual beliefs. There is no step which he is prepared to call rational from the factual supporting reasons for an ethical conclusion, to that conclusion. The beliefs cause the attitudes in as non-rational a way as that in which wind drives sails. Stevenson is also anxious that we should not dismiss with contempt the more completely non-rational aspects of ethical discussion in which we affect each other's ethical attitudes by the use of emotive language or influence each other's emotions in other ways which do not require the mediation of rationally based changes in belief. Admittedly, he says that he is describing ethical discussion as it is, rather than as it ought to be. To say what it ought to be would be to express attitudes of his own, which he does not regard as the task of the kind of philosopher like himself who is, in his professional work, trying to understand ethical discussion rather than to participate in it. However, he does indicate that he would not approve of attempts to make ethical debate entirely a rational affair even to the extent that that is possible.

But surely Stevenson greatly underestimates the place which the attitude theory can allow for rationality in ethics of much the same sort as is thought desirable in science, history or philosophy. For to that great extent to which such rationality is a matter of seeking for logical consistency, and for some more general coherence in the total spread of one's opinions, it can have full play in ethical thought, conceived along attitudinist lines. All that is missing is reasoning of an inductive or abductive sort (in the parts of ethical thought which cannot be hived off as 'factual') and even for this there is a kind of substitute.

Certainly the attitudinist can allow *deductive* reasoning full sway in ethics, since attitudes can be logically inconsistent one with another just as much as can beliefs. Just as two beliefs can be such that they cannot both be true, so two attitudes can involve wishes that cannot both be realized. Disfavouring abortion in all cases is inconsistent with favouring it in some, since there are circumstances in which these attitudes will prompt attempts to permit and to prevent the very same act.

Consider a simple example of the power of deductive reasoning in ethics. Let us suppose that an acquaintance has said that Smith acted wrongly in telling a fairly minor lie in order to promote his career, and suppose that a little later he himself tells a fairly minor lie in order to promote his career. On finding his lie out, one challenges him, and asks whether he thinks he acted wrongly, and he denies this, excusing himself on the grounds that no one was much harmed. One then reminds him of the Smith case, and asks whether he still thinks that Smith was wrong. The chances are that he will see his inconsistency and revise his moral views, either exonerating Smith or condemning himself. If he does not, he has in effect accepted at one time the proposition (the matter put upon the mat) 'Telling even a minor lie to advance one's career is always wrong, even though it harms no one' and at another time 'Telling a minor lie, which harms no one, in order to advance one's career is not always wrong'. If he refuses to admit subsequently that on one of these occasions his moral judgement was mistaken, he is certainly contradicting himself and thus being irrational. It may be suggested that he may simply qualify somewhat and say that it is wrong for others to tell such lies and not for him, and that there is no purely logical inconsistency here. Even so the qualification has been imposed by deductive reasoning. Besides, few would be prepared consciously to adopt this attitude, and would be concerned to square their final position with maintenance of an attitude less partial to themselves.

It may be asked what there is to stop people adopting inconsistent attitudes. If one asks why it is bad to hold logically inconsistent factual beliefs, the most obvious part of the answer is that if one does so, one guarantees that one will have a false belief within this area of one's thought. Since the whole point of belief is to be true, logical inconsistency in belief defeats the aim of belief. It appears that logical inconsistency as between ethical beliefs (which, according to the attitudinist, are really attitudes) cannot be condemned on these grounds, since there is no truth (in the deep sense in which truth is a genuine thing to be pursued) in the offing. However, one can say that a logically inconsistent set of attitudes bears the inevitable seeds of emotional disharmony within the individual and is likely to make effective action impossible by promoting patterns of behaviour which prevent each others' successful issue. If one really disfavours all lying, one will disfavour one's own lying, and if this is combined with an attitude of favouring one's own lying in certain cases, one will at once favour and disfavour certain possible lines of action on one's own part. This will cause emotional disharmony and make it difficult to act effectively in any way at all. Doesn't this conflict with the basic aim of emotion, and action, of reaching a general stance in and towards the world with which one can be content, just as inconsistency in belief conflicts with the basic aim of belief, namely truth?

Perhaps it will be suggested that there is an incoherence in setting out to have partly false beliefs, but that there is no similar incoherence in setting out to have emotional conflict and even a certain amount of chaos in one's behaviour. As against this, it may be said that it is psychologically impossible (or perhaps impossible in some stronger logical sense) to retain either beliefs or attitudes of which the inconsistency has become manifest to one.

However that may be, it is, surely, part of the very meaning of being rational that one tries to organise one's mental stance towards the world so that it is consistent and comprehensive, consistent in that its elements do not frustrate each other, comprehensive in that it covers one's stance to as wide as possible a range of phenomena. The mental stance the rational person seeks to organize in this way includes both beliefs and attitudes. Beliefs are inconsistent when they *cannot* both be true, attitudes are inconsistent when there are circumstances in which they *cannot* (in the same sense of 'cannot') both be fully actualised in feeling and action.

Attitude, then, can be inconsistent with attitude, and ethical statement with ethical statement. But what of attitude with belief, and ethical with factual statement? Well, certainly an ethical proposition which incorporates or presupposes a factual claim can be inconsistent with a factual statement. 'He was wrong to steal that money' is logically inconsistent with 'He did not steal the money'. However, according to the attitudinist, nothing merely factual can be formally inconsistent with a pure ethical statement like 'It would be wrong to steal that money'.

In this intuitionism and attitudinism are at one. Both insist on a strong fact/value distinction, according to which there are pure ethical statements which are logically independent of all factual statements. The reason for this, according to the attitudinist, is the distinction between belief, which sets out to represent how things are, and attitude, which is rather a matter of how one wishes (not merely idly, but in a way which can produce appropriate action) them to be. However, that is not in the least to deny the importance and validity of concatenated pieces of reasoning in which both factual and ethical premisses combine. The typical case is one in which a factual statement points out that a situation is one of the kind in which an ethical statement asserts that a certain obligation holds, that is, one in which it has expressed the wish that people should act in a certain way or meet the disfavour of others.

If the attitudinist can make sense of deductive reasoning as applied to ethical statements, it seems that he can make sense of the embedding of ethical sentences in complex sentences where the attitude they would express on their own is, so to speak, held in reserve. Consider the following discourse. 'If polygamists are bad men, Solomon was a bad man. But, as our faith assures us, Solomon was not a bad man. Therefore, polygamists are not, at least in all cases, bad men.' The antecedent and consequent of the first sentence do not, in this context, function to express disapproval either of polygamists or of Solomon. However, for the attitudinist, they function in a perfectly intelligible way in a discourse which explores relations of consistency and inconsistency between various possible and actual attitudes.

This is important, because it is sometimes given as an objection to attitude theories that they cannot make sense of subordinate ethical clauses. Even this one example suggests that this objection has little to it. Perhaps it has more force against emotivism than against the attitudinism I have described. Stevenson himself thought that it was only in virtue of a descriptive meaning, which ethical sentences typically had, as well as their emotive meaning, that they could function in this sort of context. He may have ceded too much even on his own theory, but certainly there is no problem here for attitudinism divorced from emotivism.*

* It is true that the way in which ethical statements occur embedded in such contexts as the above has to be explained at the pragmatic rather than the semantic level. That is, it has to be understood in terms not of relations between what statements say but between what they express. But it is most doubtful that

However, when it is thought that attitudinism renders ethics irrational, the usual point is that ethical conclusions cannot be inferred in any rational manner from premisses which are purely factual. That means that they cannot be supported, or more importantly tested, by observation. However much you pile up facts about the world in order to support an ethical case, you cannot get an ethical conclusion unless you take some ethical premiss for granted, in a way which usually just begs the question against those with whom you disagree. Thus if people are trying to decide whether the use of human foetuses for experimentation, and perhaps their deliberate production for this purpose, is morally acceptable, no amount of factual knowledge will settle the matter of itself. If it seems that things could be finally settled by deciding whether a foetus is a person, this can only be because 'person' is used to ascribe a moral status rather than merely to

this seriously sets the logic of discourse involving ethical statements apart from purely factual discourse, since it is hard not to think that logical relations, even when explicable at a purely semantic level, are ultimately rooted in pragmatics. Thus I see no need for the attitudinist to move to the quasi-realism of Simon Blackburn in order to allow for the occurrence of ethical sentences as subordinate clauses. (See *Spreading the Word* (Oxford, 1984) chapter six.) It is true that the attitudinist thinks the form of ethical statements a little misleading in that they look as though they had a semantic content in the sense in which factual statements do. But in explaining philosophically what is going on, the attitudinist need not, as Blackburn seems to think, treat them as though they were a kind of bogus factual statement. There is something to be said for a *phenomenological* 'projectivism' if it claims that to have emotions and wishes is to experience the world as having certain qualities which are only there for one's own consciousness, or others like it, but in the absence of such claims it is not clear why it is an advance on a sophisticated attitudinism. Perhaps the best reason given in favour of quasi-realism, rather than straight attitudinism, is that it does justice to the way in which we seek, and in which the philosophical commentator may favour our seeking, a unified and agreed total stance in ethics of a kind which has its least problematic exemplars in types of enquiry where genuinely objective truth is sought, and which can be carried over from there to the ethical case where no genuine truth is in the offing. There is something in this. Still, if there is no real truth in ethical claims, at least in their distinctively valuational aspect, it may be thought the part of honesty to recognize (and not just when philosophising) the truth of the attitudinist account, and recognize that there is not the same ground for expecting convergence here as there is wherever a more full blooded realism is in order. Stevenson, indeed, says that it is possible that all disagreement in attitude will be resolved when disagreement in belief is so, but that there is no logical guarantee of this, and thus no guarantee that rational methods, or perhaps any other, will tend to produce convergence.

describe. ('Metaphysical' facts about the presence of a soul would be equally incapable of closing the matter ethically.) Attitudinist and intuitionist agree that this could only be denied through the naturalistic fallacy. Of course, intuitionists and emotivists disagree about the availability of 'rational intuition' (guided by such principles as that of organic unities) but so far as the relation between the merely factual and the ethical goes they are at one. Yet this is only the corollary of the view that there is a sharp distinction between ethical and factual claims, and hardly justifies calling either theory irrationalist.

Perhaps the charge that attitudinism makes ethics peculiarly irrational is merely a way of saying that it denies that there is such a thing as objective ethical truth. However, it is better to distinguish the issue of objective truth from that of rationality. Rationality can very properly be specified not in terms exclusively of methods supposed to lead to truth but more generally in terms of methods for reaching a consistent and comprehensive stance towards the world as it really is, something perfectly possible in ethical thought as the attitudinist describes it. It is, indeed, an implication of attitudinism that induction and abduction cannot be invoked in support of ethical statements as they can in support of factual ones. However, even if induction and abduction are not available here, something quite similar is, namely the moving tentatively to general conclusions on the basis of one's responses to particular cases, and the testing of general conclusions by how acceptable one finds, in practice, the responses to particular situations which they dictate. I am gesturing here towards the so called method of reflective equilibrium whereby general ethical principles and particular ethical judgements, as they arise fairly spontaneously, are each reconsidered until a consistent position is found with which one can live. (See RAWLS p. 48 et ff.) Thus in general it seems a mistake to see attitudinism as the enemy of reason in ethics.

6. It is time now to consider more precisely what the attitudinist says about the actual meaning of typically ethical words, like 'good' 'ought' 'right' and 'wrong'.

Let us say that the meaning of a statement is cognitive if and only if there is a certain belief such that one is speaking either insincerely or incorrectly if one makes that statement or assents to it as uttered by another without having that belief. And let us say that the meaning of a statement is valuational if and only if there is a certain attitude to which it is related just as a cognitive statement is related to a certain belief. It should be emphasised that the 'correctly' which is negated by 'incorrectly' in the above means *with linguistic propriety*. One speaks with linguistic propriety and sincerity when saying 'It will rain this afternoon' if and only if one believes it will; the truth of the matter has nothing to do with it. The distinction between cognitive and valuational meaning follows naturally from our account of the way in which statements express beliefs or attitudes. (I talk of 'valuational' rather than evaluative meaning, as it seems improper to capture 'evaluative' for some special theory. No one would deny that statements about what is good are evaluative.)

It will be clear that – according to the theory I am expounding – the meaning of the more basic sort of ethical statement is valuational rather than cognitive. However, the theory allows that a statement can express both a belief and an attitude. For example, 'He is a wicked old man' expresses both the belief that he is an old man and some kind of disapproval of him. It should be noted, however, that a statement like 'If you do that, you will be acting wrongly' has a purely valuational meaning. It expresses one's readiness to have a certain attitude to the person named by 'you' if he acts in a certain way, and in a broad sense this readiness is itself an attitude.

We can now say that the meaning of an individual word is valuational to the extent that its prime role is to make the statements in which it occurs express certain attitudes, and that it is descriptive if its prime role is to specify the content either of the belief or of the attitude which sentences in which it occurs express. By the content of an attitude is meant what it is an attitude towards, while the content of a belief is the state of affairs the reality of which it affirms.

The main thesis of the attitudinist theory of ethics can now be expressed as the claim that the most general ethical words have a primarily valuational meaning, while to be properly ethical at all, a word must have a meaning which is partly valuational.* Thus

* A more complete account would have to consider how these concepts apply in relation to such non-declarative sentences as imperatives and interrogatives. It

'good', 'bad', 'ought', 'right', 'wrong' have primarily valuational meanings, at least in a properly ethical use of them, while words like 'brave', 'lazy', 'rude', 'selfish', 'considerate', and so forth which have a much more settled descriptive meaning, have a partly valuational meaning. It need not be insisted that all valuational statements and words are functioning ethically, for there may be non-ethical attitudes.

In the case of words which combine a fairly definite descriptive meaning with a valuational meaning it is rather a puzzle to say what correct linguistic usage bids one do, if one recognizes that something answers to the descriptive meaning, but does not have the attitude towards it which the word expresses in virtue of its 'value charge', as one might put it. Take the word 'loyal' and consider someone who deplores the kind of character it indicates. Does he say of someone who does not have this character that he is 'not loyal', in order to express lack of approval of him, or does he say that he is loyal, but seek to negate the usual valuational meaning of the word, by inserting some adverb like 'deplorably' before it? We need hardly answer this – it depends upon circumstances. The important thing to note is that such words will only have a value charge as part of their standard meaning, if the users of the language mostly have a certain shared attitude to what answers to the descriptive meaning. Thus the person who faces this problem is, in the nature of the case, unusual. The value charge of such a word reflects the *mores* of the language users, and they may be unaware that it has two distinguishable components to its meaning. The reflective analyst of language, however, will see that words are playing this double role and how their meaning thereby incorporates the specific values of just one possible type of society.

It is sometimes said that if one wishes to reject such values one must do so by rejecting the very concepts incorporated in such words, and cannot do this merely by applying their negation. It is

would also need to explain why I have spoken of 'cognitive' meaning in connection with the meaning of a statement and 'descriptive' in connection with the meaning of a word. It was an unsatisfactory aspect of Stevenson's contrast between descriptive and emotive meaning that he classified a word like 'good' as emotive and a word like 'cat' as descriptive, and associated this with the distinction between attitudes and beliefs, although 'cat' is just as suited to occur in a statement expressing an attitude to cats as in one expressing a belief about them.

also sometimes supposed that their existence shows that the strong fact/value distinction made by attitudinists and intuitionists is misconceived and that in our language, as it stands, factual premisses can logically imply evaluative conclusions. Thus a true statement of the facts of the case can logically imply a statement ascribing the ethical predicate 'dishonest' to someone. However, the attitudinist will say that the only real deduction is at the descriptive level, and that the distinctively ethical part of the conclusion springs from the value charge of the word which expresses an attitude from which one can disassociate oneself without being in the least irrational. Within the limits of his unsatisfactory emotivist position Stevenson deals with all this very effectively in what he calls his 'second pattern of analysis' of ethical words and the same main points can be made in terms of valuational and descriptive meaning. A related point requiring emphasis is that much of what one is inclined to regard as the mere 'facts of the case' are often 'institutional' facts our recognition of which is itself partly the adoption of an attitude.

It has been suggested that non-cognitive theories of ethics do best with those ethical words such as 'good' and 'ought' which are most plausibly represented as purely valuational, while cognitivist theories (which take moral knowledge and truth seriously) do best with words such as 'brave' 'loyal' and so forth (LOVIBOND). This seems wrong. It would be true at most of a certain sort of naturalistic cognitive theory, for it is certainly not true of Moore's cognitivism. Moreover, Stevenson's theory, and the attitudinism sketched above, are at their most convincing in their treatment of value charged descriptive words, among which indeed even such words as 'good' can be counted when we are concerned with the meaning that they have in a homogeneous society. (A suggestion made by Alasdair MacIntyre that emotivism is promising only as an account of the use of ethical words in a society lacking shared values could not reasonably be extended to the attitudinism I have described.)

I have already suggested that Moore's doctrine of the naturalistic fallacy is one of the inspirations for attitudinism. That is why attitudinists are sometimes rather oddly lumped together with intuitionists as non-naturalists, in spite of the fact that Stevenson. who saw himself as bringing ethics down to earth, called his view a form of ethical naturalism. Whatever the right labels, the attitudinist agrees that no properly ethical expression can be adequately defined in purely naturalistic or even metaphysical terms but offers what he thinks a better explanation of this fact than the invocation of non-natural properties, namely that such definitions ignore valuational or emotive meaning. This is thought a better explanation both because it avoids the metaphysical extravagance of non-natural properties, and because it clarifies the relation between moral judgement and action. For if 'good' is the label for a property rather than the expression of an impulse why should noting its presence have anything more essential to do with an inclination to promote it, than noting that a wall-paper is blue has to do with choosing it?

But although the attitudinist agrees with the intuitionist that the meaning of ethical words cannot be exhaustively analysed in naturalistic or metaphysical terms he takes a more positive view of the kinds of definition which Moore was so concerned to refute, for he sees them as examples of a particular type of definition, which has a legitimate place in discourse. Stevenson christened them 'persuasive definitions'. Such a definition takes a word with a certain valuational meaning, and rather vague descriptive meaning, and broadens, narrows or shifts the descriptive meaning so that the emotions expressed by the valuational meaning are directed at more or less substantially different targets. Thus I may define 'propaganda', which usually has an unfavourable valuational meaning, so that it refers to any vigorous attempt to change people's outlook. As a result various things, such as missionary work, now fall under a word which directs disfavour at them.

There is much that is illuminating in Stevenson's treatment of persuasive definitions, especially if we disentangle the essential points from the unsatisfactory notion of emotive meaning.* What he tends to ignore, however, is what one might call value charged definitions which do not function as persuasive devices for modifying attitudes but serve simply to formulate, and perhaps

* Subsequent discussions of the relation between facts and values would often have benefited by reference to his work here. So would some discussions of the status of such questions as 'What is law?' as, for instance, in chapter two of Ronald Dworkin's *The Empire of Law*. That it oversimplifies is no justification for ignoring it.

endorse, combinations of descriptive and valuational meaning in ordinary usage which reflect widely shared attitudes. In any fairly homogeneous society the meaning of even such basic valuational words as 'right' and 'wrong' can only be given in such value charged definitions, which point to the society's shared values. In that society the rightness or wrongness of an action does in a sense simply follow from the facts of the case, but only because its language incorporates values which the attitudinist will see as rationally detachable from beliefs of a genuinely factual nature.

As will be seen later, I do not myself go all the way with the attitudinist theory of ethics. But it has real merits, illuminating much that other theories leave obscure. As for the role it ascribes to reason, Stevenson himself played this down unnecessarily. Properly developed, it gives reason, conceived as the quest for coherence, comprehensiveness and sensitivity to reality, almost as great a role as it can play anywhere. What it does, however, suppose absent in ethical deliberation is any actual truth of the matter at which it aims. Here it distinguishes it from factual enquiry, at least as that is conceived by those with a robust sense that there is a way things really are in the world. It may see this as meaning that reason has less hope of producing convergence in ethics than on factual matters.

It should be noted that the attitudinist theory of ethics operates at a rather different level from a theory such as utilitarianism. Utilitarianism is a properly ethical theory, inasmuch as it puts forward the basics of a moral position, while emotivism belongs to what is sometimes called meta-ethics which professes to explain the nature of what is going on when an ethical or moral view is advocated. (We saw in Chapter One how some thinkers have advocated utilitarianism in a consciously attitudinist spirit.) At one time the question was much debated whether meta-ethics can or ought to be neutral as between ethical theories, or indeed whether the two should be distinguished at all. The truth seems to be that this question itself cannot be answered in a way which is neutral between different positions in meta-ethics and ethics and is hardly worth discussing in the abstract. For that reason I shall not attempt some neat separation of the ethical theories which I discuss into these two classes.

7. I conclude with some remarks about various points of view in

moral philosophy which bear on the claims of the attitude theory.

(1) There is a tradition in Austrian and German philosophy which maintains that just as a judgement can be true or false, so can an emotional attitude be correct or incorrect. Moral philosophers who hold this sort of view are wont to agree with the attitudinist that moral judgements, or value judgements on which they are based, express emotional attitudes, but will insist that these can be correct or incorrect in quite as objective a way as that in which judgements can be true or false. Chief representatives of this view are Brentano and Meinong. The attitudinist would say that calling an emotional attitude correct or incorrect is simply expressing a higher order emotional attitude towards it.

(2) Modern critics of an attitude theory have often argued, not unconvincingly, that the methods of reasoning used in support of ethical views, or of trying to reach agreement on them by a fair minded exchange of views between reasonable persons, neither need be, nor commonly are, different from those used in factual matters. A good exponent of this approach is Renford Bambrough. This seems a fair point to make against some of Stevenson's and Ayer's contentions, but the attitudinist, as I have characterised him, in this chapter, would acknowledge Bambrough's points, yet still think it important to insist on the fundamental difference between trying to decide what is the case and trying to decide what one would wish to be (and will try to make) the case and on associating ethical claims with the latter.

(3) There is a modern attempt to defend what is called objectivism in ethics which turns on viewing both facts and values as essentially social constructs which are what they are because they are either what society drills us into accepting or what exceptional individuals can persuade us follow from the procedures we have been drilled into accepting as correct. A subtle exponent of this approach is Sabina Lovibond. Stevenson would certainly respond by insisting that however much our conception of the world is socially determined, and even if there is no level of brute fact which is just how things are, there is still all the difference between a socially agreed mode of representing actual or expected, but not necessarily welcomed, reality, and an effort to decide on what to favour and what disfavour in this reality.

(4) A view which has evoked a fair amount of discussion is 'the

error theory of ethics' put forward by J.L. Mackie. A brief discussion of it will lead naturally also to some remarks on some views about ethics put forward quite recently by John McDowell.

Mackie distinguishes four types of view on the status of value and obligations, (1) non-naturalist objectivism, (2) attitudinism, (3) modern naturalism, (4) error theory, and comments on them somewhat as follows.

The first view makes values and obligations part of the fabric of the world. However, if they are to play their required role for ethical thinking, these values have to have an intrinsically prescriptive character, so that to know them is necessarily to have the will affected in a certain way. Attitudinism directly relates ethics to the will, but it neglects the phenomenological fact that we think of values and obligations as something actually there. A certain sort of modern naturalism takes ethical expressions as standing for various natural or empirical features of the real world. (One well known exponent of such a naturalism is Philippa Foot.) These are ones in which humans take a special interest. However, this neglects the categorical or unconditional prescriptivity which we think pertains to values, for it makes the relevance of ethical truths to us turn on contingent facts about the will. Mackie's own theory of ethics is (or includes) an 'error theory'. We ordinarily think of values and obligations being really there, but we also think of them as intrinsically prescriptive, that is, such that one who knows them must be affected conatively in an appropriate way. Mackie thinks the idea that there could be such inherently prescriptive properties and relations, which are part of the genuine fabric of the world, too queer and incoherent to be true. So he concludes that ordinary moral judgements express the erroneous view that there are objective features of the world which intrinsically (and not merely because, as a matter of contingent fact, we respond to them in certain ways) require something of us. Thus Mackie is what he calls a 'moral sceptic', though he is in favour of continuing to live in the light of moral values and obligations which we acknowledge as a human invention or construction serving widely shared human purposes.

John McDowell has challenged Mackie's assumption that a genuinely objective feature of the world must be something whose thereness in the world has nothing to do with human responses to it. He notes that Mackie compares values to colours, and describes both as things which are normally thought of as existing independently (and which conceivably might have done so) but which a reasonable scientific view of the world exhibits as not really there. McDowell objects that we should not think of what is really there in the world as something whose thereness has nothing to do with our responses to it. In fact, he claims, colours present themselves as what they truly are, features of the world which are intrinsically to do with how we see them. The case is quite similar with values, which present themselves as real features present in the world which are intrinsically *to do* with how we respond to them.

To recognize the value present in a situation (he urges) is not merely to have an attitude which someone else who conceives the 'factual character' of the situation in exactly the same way might lack, but to conceive it in a particular kind of way which could not be duplicated in someone not thus drawn to it. But that is only a reason for saying that the value is not really there in the world if we presuppose a scientistic view of reality for which it is of itself necessarily 'motivationally inert' and cognizable in a manner which has nothing essentially to do with being attracted or repelled by it. The real world of 'scientism' which could be characterised in terms having nothing to do with our particular ways of responding to it is a myth, with which it is quite inept to confuse the real 'real world'.

The position I shall be advocating in part two has certain affinities with aspects of each of these views, yet each seems unsatisfactory as it stands. I quite agree with Mackie that colour presents itself as being as much part of the fabric of the world as, say, shape and that much the same is true of value, and also agree that they seem therefore initially intelligible as being there in a manner not intrinsically bound up with our responses to them. He is surely also right that philosophical reflection gives adequate ground for denying that colour and value can be there independently of the consciousness of them. But this is only a ground for drawing a strong contrast between them and 'primary qualities' like shape, for those who, like Mackie, take a strong realist line about the latter which I find unacceptable.

But even without challenging such realism, I would argue that values, by being made dependent on consciousness, do not cease to be part of the fabric of the world. For consciousness is itself part of that fabric, and, or so I shall argue in Part Two, the value features which are found there are real properties of certain experiences (and indeed of the physical and social environment in our personal versions of it) which have an intrinsic prescriptivity. However, I shall partially echo Mackie in insisting that many values are only there as part of a shared social construction. Still, even these, to whatever extent they are not merely believed in, but have come to be actually felt in the world as we experience it, will thereby have their place in the real fabric of the world, which must certainly include all personal or socially shared versions of it.*

McDowell's excessively subtle thesis teeters on the brink of the view that 'This is red' means 'This looks red to standard observers' and 'This is good' means 'This appeals in such and such a way to a normal person'. However, he rejects this view on the ground that although the redness of a red object is not something which merely happens to look red to us, since a redness which did not look red would not be redness, nonetheless 'red' does not mean 'looks red' since only one who understands 'red' can know what 'looks red' means. Thus redness, rather than merely looking red, is basic but its being there is bound up in a way which is manifest from the start with how it looks to us. Similarly the value of something is intrinsically bound up with the way in which someone who recognizes it is drawn to it, or repelled by it if the value is negative, but is not merely a disposition to attract or repel, for we cannot be thus attracted or repelled except by recognizing (or at least seeming to recognize) a value.

I agree with McDowell that 'red' is more basic than 'looks red',

* There is a certain obscurity as to what the error is which the error theory ascribes to ordinary moral and value judgements. If it is an error about what ordinary ethical judgements mean then that is not itself enough to justify saying that these ordinary judgements themselves incorporate an error. If it is an error incorporated within them, that suggests that these judgements envisage certain definite properties which in fact nothing whatever possesses. But if the idea of there being value properties is as incoherent as Mackie thinks it, how did we ever come to envisage not just the idea of there being properties which answer to some general description, but certain definite such properties? Could we have the idea of a certain definite property if there are no properties at all of the genus to which it belongs, nor any elements out of which the idea of it could be constructed?

but would explain this in a way which is nearer to Mackie's outlook. 'Red', in the most basic sense, refers to a quality whose true locus is only in the perceptual fields of beings such as ourselves, but which in our most basic conceptualisations of the world we think of as being present more stably in a real world the character of which can clash with the world as immediately presented. Thus we fabricate a deeper sense of being red in which an object *is red* only if it is red in a stable fashion in the perceptual fields of all observers and which is therefore treated as part of the real world which exists independently of us. In contrast with this fabricated 'real' redness the redness which immediately appears and from which the whole conception of red is obtained is downgraded to something which pertains only to how things look to us. I shall be suggesting in Part Two that the basic idea of 'good' and 'bad', and of all value, arises similarly from what feels good or bad to us, and that here again this good and bad which is demoted to the role of what only feels so is ultimately the value which is really there in the world, besides which the good and bad of our social construction of reality is only a kind of useful fiction.

When I develop that thesis it will be seen that I agree with McDowell as to the intrinsic attractiveness or repulsiveness of value qualities. However, I will not, as McDowell comes too near doing, assimilate the recognition of the existence of a value with the being appropriately drawn to it. I will contend rather that the judgement that it is there has a necessary effect upon how we are inclined to act. In this respect my position will amount to thinking that there are intrinsically prescriptive features to reality, but this will not be done by blurring the distinction between judgements about how the world is characterised and the ways in which we respond to it, as with McDowell. Value properties, for me, are really there in the world quite apart from a valuer's responses to it, but have a necessary effect on the will of him who knows of them. These value properties are, indeed, only really there in the world insofar as they qualify states of, or immediate presentations to, consciousness, since they consist in the various ways in which things feel good or bad to sentient beings. The extent, however, to which that marks them off from what are typically thought of as more 'objective' features of the world in which we live depends on how far one thinks that has a character independent of any form of sentient experience.

CHAPTER IV

Some Great Historical Moralists

In this chapter I shall be considering some of the great historical Western moral philosophers of the past, confining myself, however, to the period of modern philosophy, which is usually conceived of as starting in the seventeenth century. (I regret the omission of ancient philosophy, but that could not be decently treated without doubling the historical part of this book.) The treatment will be unavoidably brief. It should serve the purpose, however, of providing a sense or a reminder of something of the range of alternative answers there are to questions about the rational foundations of ethics.

Spinoza

Baruch or Benedict Spinoza (1632–77) attempts to give ethics an unequivocally rational foundation. He puts forward an ethic consisting of habits of mind and of behaviour to which he thinks one will inevitably move to the extent that one has rational insight into the human situation and is under the control of that rational part of one's nature which gives one unity as a personality. In some respects his ethics is a representative in the 'modern' world of the same type of approach as was taken by Plato and Aristotle. For Spinoza the good life is (in effect) one in which human beings fulfil their essential natures. It is true that Spinoza is famous for having denied the existence of final causes in nature in the sense in which Aristotle believed in them, that is purposes for which each thing existed. However, his view that each thing has an essence and that all its active behaviour is to be explained by its inbuilt *conatus* or striving to preserve its own essence has a strong Aristotelian flavour. What he is mainly denying when he denies that there are final causes in nature is that the existence of each individual sort of thing is to be explained by its serving some cause beyond it, in particular some kind of human interest. For he is above all concerned to deny that the point of the universe is somehow to serve human interests. Rather it is a spiritual (as well as physical) unity which simply is what it is, and in which human beings exist as what they are, without itself having a point beyond itself or the things within it having a point beyond themselves. This is no time to go into further consideration of how Spinoza's view of the world may relate to the Aristotelian. I simply note that in some respects we are, by dealing with Spinoza, covering a type of ethics which our neglect of ancient philosophy might otherwise have led us to omit. This seems to be true in spite of the fact that Spinoza was very much of a generation which was concerned to dissociate itself from the Greek inheritance, and indeed he represents something of a fresh injection of Jewish moral feeling into the main Christian current of Western thought. His thought also has elements of the Stoicism of Athens and later of Rome.

Spinoza's ethics is closely related to his metaphysical views. Indeed, the essential statements both of his metaphysics and of his ethics are presented in one work called simply *Ethics* (*Ethica* in the original Latin in which it was written).

Spinoza was a Dutch Jew of Portuguese descent. He was excommunicated in his early manhood from his synagogue on account of his unorthodox opinions. Later he became associated with a Christian denomination known as the Collegiants. His metaphysics is pantheistic. Nature and God are identified, and are conceived as an infinite reality which exists both as an all comprehensive mind and as an infinitely extended physical system, and in other ways unknown to us, all these being different aspects of one and the same single reality. Our bodies are elements in this infinite physical system and our minds are elements in the all-comprehensive mind, and thus aspects of one and the same individual fragment of the total divine reality.

Each individual fragment of the total reality has an individual essence and its behaviour is entirely explicable as its effort to keep itself in being in as full a fashion as its environment allows. More strictly, this effort is simply that essence itself which precisely is its effort to keep itself in existence in its own distinctive way. Thus human beings, animals, plants, and all other genuine units in the world, on whatever scale, struggle each to exist in its own way, this, on the one side, being a matter of a certain physical individual having an inbuilt tendency to operate, so far as is possible, to preserve itself, in its distinctive form, and on the other hand, of the corresponding mind having an inbuilt tendency to preserve its particular form of experience. Each human being strives to keep himself or herself going in their own particular character, but being made up of parts, those parts which have any kind of individuality also try to keep themselves going. Thus are we victims of various sorts of obsession when certain parts of ourselves seek to preserve themselves even at the expense of the whole personality. So far as the whole personality does not fall victim to the pressures either of the external environment or what one might call the internal environment of its own parts – in the form of various lusts and obsessions – it achieves its only conceivable goal, that of prosperous possession, in the fullest manner, of its own nature.

This may sound like a purely mechanical process in which the most powerful essence wins, without there being any room for ethical or indeed any other decision. But we now have to reckon with the fact that the peculiar essence of a human being is rationality, that is, each human essence is some particular form of rationality. (Some of what Spinoza says would imply that every unit, even a plant or atom, is on its mental side a particular sort of rationality. It is the quest to think of that peculiar essence of which its physical nature is the more or less full actualisation. However, although this is in a manner so, 'rationality' in a more usual sense refers to the particular sort of rationality distinctive of human beings, by which they participate in their own way in the ultimate rationality of the universe.)

As a particular form of rationality a human being, considered as a total personality, is above all an effort to think things out by having a firm conception of its own nature, as this exists in the physical world, and of what will assist this nature to keep in existence. This effort to keep a clear conception of its own nature before it is the mental side of the human psyche when it genuinely acts in the world and is not merely subject to external or internal forces. Thus the main task of childhood is to achieve a sense of its own distinctive personal identity and to do what will reinforce it. (This reference to the task of childhood is rather my own effort to develop Spinoza's own richly fruitful but rather abstract ideas in more concrete terms than a direct report upon what he says – the same goes for some other of my remarks.) This essentially rational nature of the human essence explains why thought – such thought as Spinoza tries to express in his work – can influence our behaviour, so that we are guided by conceptions of good and evil. 'Good' in this context means what promotes our distinctive mode of being and 'evil' what hinders it. If we think that there is any objective or independent good or evil in the world, we are mistaken. Ultimately the universe in its totality is in some sense perfect, as also are its elements understood as required by the overall nature of that totality, but from the point of view of us who struggle within it things are called good or bad (or by some such words) according as to whether they promote our particular struggle to actualise our own nature.

The fact that a particular sort of rationality is of our very essence explains why rational thinking can lead us to ethical self control. We simply cannot form a clear and distinct idea of the fact that something is conducive or inimical to the actualisation of our essences without this influencing our behaviour towards or away from it. There are plenty of interfering factors but these are inevitably to some extent weakened by a rational understanding of them, at least when this engages appropriately with our basic *conatus* (or effort to preserve our own nature) and thereby acquires the requisite emotional power.

Part of what we are striving for as a necessity of our being is simply survival in physical and mental health. Simply to live in the world performing one's proper biological and mental functions, eating, engaging in physically healthy activities,* appreciating the beauties of nature and of art, trying to understand the way things work, developing one's skills is joyous

* Spinoza's actual personality and background seem to have made him very conscious of the dangers of excessive sexual desire, but upon the whole his thought is sharply critical of extreme asceticism, though he insists that bodily appetite should be under the control of the main good for humanity, that of rational thought. Spinoza's ideas often point beyond the development they actually receive at his hands, and really seem to indicate that conscious creative activity, not necessarily of what is typically called an intellectual kind, is the completest fulfilment of our essence.

and an end in itself. When this is not enjoyed it is because our own essential nature is frustrated in its enjoyment of itself through physical or mental disease or by physical or social circumstances. Spinoza offers us no magical solution to these problems. Hard thinking and research may aid us individually, and eventually as a species, to counter some of these problems. Yet we must also simply learn to accept our status as mere fragments of total nature who cannot help sometimes being overcome by forces we cannot control, whether of a physical or of a psychological nature, or by the the assaults of irrational men. Thus we must strike a fine balance between simply learning to put up with things through realizing the inevitable limitations on satisfaction in the life of a finite creature, and between vigorous efforts to put the situation right. However, even if things go badly, so long as we survive as partly rational creatures we should always be able to find some satisfaction in understanding our situation and doing what reason points to as the most effective way of dealing with it. True, at death our essence is finally overcome so far as its actualisation in space and time goes, but as long as we live we can go on realizing it in as full a form as circumstances allow. (Spinoza did, indeed, conceive of us as possessing a very mysterious kind of immortality as eternal elements in the divine consciousness, but he insists strongly that this has no bearings on how we should conceive the good life.)

Spinoza was revolutionary in thinking of mental illness as the main cause of human suffering and in challenging the distinction between mental illness and vice. For him all vices are forms of mental illness, for vicious action is always a matter of submission to impulses which are seeking their own satisfaction at the expense of the satisfaction of the whole. Spinoza holds that the more we understand why we have these impulses, as also why certain sorts of thought overwhelm us with depression, the more we will get on top of them. When we form a clear and distinct idea of the irrational sides of our nature, they are transformed and become in a soberer form part of the patterned plan of our life. (He is often regarded as having anticipated elements of psycho-analysis in what he said on this.)

Thus a large part of Spinoza's message is that the point of life is simply to develop one's own powers and enjoy their exercise. However, he does see the life of one who really understands his situation as giving a central place to what he calls 'the intellectual love of God'. Essentially this is a matter of love for the total cosmic reality of which we are a part as something whose magnificence transcends our own puny being, and gratitude to it for having brought us forth in the heart of it. The more fully we have developed ourselves the more cause for such gratitude we will have, and the more we understand the cosmos, particularly by grasping the true nature of detailed parts of it and their place in the total scheme, the more we will appreciate the sheer wonderfulness of it, and arrive at a kind of mystical adoration of it.

Where does this leave the more ordinary requirements of morality? If our one inevitable aim is the enjoyment of our own powers (an enjoyment which reaches its highest level in a sense of oneness with the cosmos as a whole) what becomes of our duties to others, of keeping contracts, of helping the needy, of unselfishness in daily life, and so forth?

Spinoza harks back to ancient philosophy, and away from the Christian tradition, in seeing the virtues which ethics seeks to inculcate as essentially the qualities we require if we are to have personally fulfilled lives. Thus courage and temperance, are conceived not as a restriction on our freedom but as requirements for its full development. But this is true also, according to Spinoza, of virtues which are specified primarily in terms of concern with the well being of others. This is because men cannot properly develop their powers except in co-operation and friendship with others, and because the true goods of human life, those which we need for our own personal fulfilment, are not goods in some limited supply for which we must compete, but ones which each can the better enjoy, the more others are enjoying them. Thus if my main concern is to have a vast stock of personal possessions and control over the lives of others, I can doubtless only have them at the expense of others. These, however, are not goods which will give real satisfaction. The kind of satisfaction which can be anxiety-free is that in which my own creative powers and understanding of the world around me are at a maximum and this is satisfaction which the individual can only achieve as part of a community of persons directed at these ends. For example, musical and philosophical skill – as opposed to the external rewards of reaching the 'top' in a musical or academic career, a kind of achievement which, if given central importance in our lives, will yield no abiding satisfaction, beset as we will be by worry as to what others are or are not thinking of us – can only be developed by an individual in a propitious environment where many co-operate in working at them.

Essentially, then, Spinoza supposes that the moral virtues, and life in accordance with the demands of a sensible morality, provide both the essential background, and a large part of the content, of a human life in which each of us lives their own personal life to the full. Some pleasure denying elements of conventional or religious morality are simply to be rejected as resting on superstition. (Spinoza was opposed to the more life denying aspects of the Calvinism of his time.) The kind of morality which Spinoza recommends is one which guides us on how we may live out our lives in the fullest possible exercise of our personal urges and abilities. As such, it must help us bring the chaos of our rival impulses under the control of the personality as a whole, in the pursuit of its quest for comprehensive and balanced fulfilment, and towards an effective recognition of the fact that such fulfilment requires harmonious, caring, co-operative and stable relations with others.

But what if there is someone whose own fulfilment is simply not best promoted by this sort of relationship with others? Or what if studied departures from concern for others are occasionally required in the interests of one's own fulfilment? Spinoza says that it is no mere accident that this is never so, but that it is built into the essential nature of human beings that they need these relations with others for the achievement of personal fulfilment. (If there are people beyond realizing this, they ar e beyond the influence of any rational ethic and not among those to whom this can be addressed.)

Many think it a fatal weakness in any such attempt to base morality upon egoism that this answer is false. Perhaps Spinoza could have strengthened it in various ways, by saying that people cannot on odd occasions deliberately act out of the character they try to give their lives, without destroying that character, and that for the rational person the character of a life which includes good relations with other people at large is essential for personal fulfilment. Whether it is possible or desirable to try to give ethics an egoistic basis in this way must be discussed later.

It should be emphasised, however, that although in a sense Spinoza recommends the ethical precepts he endorses to each of us as what we will accept if we act with a view to our own best interest, these best interests are conceived in a way which is very far removed from the goals of what is commonly called egoism. For the best interests are essentially those of a full and developed personality to which caring relationships with others are integral.

What is forceful in Spinoza's approach is his belief that it is mere empty verbiage to recommend conduct to people unless this engages with their real motivations. This seems eminently reasonable, quite apart from its association with the Spinozist idea that all motivation is the attempt of a personal essence to keep itself in actuality in the fullest possible way.

It will be seen that there are strong elements of an attitude theory of ethics in Spinoza. However, Spinoza, as a moralist, appeals to no merely emotive power of ethical words, nor even to facts which might serve as contingent causes of the ethical attitudes he is expressing. Rather does he base his recommendations (which, in general, are in line with the basic directives of Christianity and Judaism) on the fact (as he believes it to be) that a man who thinks adequately on the matter will see them as pointing out the means to his own deepest heart's desire of personally fulfilled living.

Eighteenth-century British moralists

Many of the most penetrating discussions of the foundations of morality are to be found among eighteenth-century British moral philosophers. The background to much of this discussion lies in the seventeenth-century English political philosopher, Thomas Hobbes (1588–1679). Hobbes had had a considerable influence on Spinoza. However, although both advanced their ethical and political ideas on the basis of the basic urge of each man to seek his own survival and welfare, Hobbes derives his from a much less lofty view of man's basic aims. Spinoza was, in any case, little attended to by the moralists of this period. He was reviled as a so-called atheist but otherwise his thought received little attention, only becoming a main force in philosophy when it was rediscovered by German philosophers in the nineteenth century. Hobbes's thought, in contrast, was both well known and widely regarded as in need of refutation on account of its low view of man.

Hobbes thought of men as having one over-riding aim, that of sheer survival. He saw the state as arising from an explicit or implicit contract among men to put themselves under a single sovereign (which could be a parliament rather than a king) which would establish peace among them. The essential reason for obeying the sovereign power was that keeping it in power was inevitably better for the security of each than the chaos of civil war, or a society without government. (Hobbes's thought was eminently practical in the time of the English civil war.) Moral principles are essentially either those rules by which men must abide if they are to unite under a sovereign power, or those which the sovereign power imposes. The reasons for obeying these moral rules are egoistic, that to disobey them is to forward the perils of social chaos, or to risk punishment.

This seemed unsatisfactory to most of the moral philosophers of the 18th century. They disliked this egoistic conception of man, and they suspected that egoistic motives might not be sufficient to underpin morality. One of the most powerful of these thinkers was Francis Hutcheson (1694–1746). He believed that there was a great fund of benevolent feeling in man which could often counteract his egoism, and that man also possessed a moral sense, to which benevolence appeared with a special and attractive quality of moral goodness. This moral goodness was not exactly out there as an actual property of the benevolent feelings and actions which presented themselves to an observer as possessing it. Rather, the moral goodness was really the power to produce a certain sort of pleasing sensation in the observer. But a quality which is really a power to produce feelings or sensations in an observer is in its way a perfectly real and objective quality. That is, after all, just what, according to the dominant philosophy of Hutcheson's time (that of John Locke) such so-called secondary qualities as colour, sound and smell are. They are not there apart from an observer, they are rather the tendency of objects to produce certain sensations in one. As to which sorts of things will produce these sensations, that has been settled for us by nature (ultimately God) in such a way that normal men will inevitably agree as to the colour and smell of things. Thus these qualities have a kind of objectivity, or at least inter-subjective validity. The same goes for moral qualities, such as the goodness of benevolence. All normal men find that the same sort of actions and feelings, basically the various forms of benevolence, possess this quality.

Although Hutcheson thought of himself as defending the reality of moral distinctions, and the genuineness of a morally good benevolence which was not egoistically based, other thinkers were not happy with his treatment. They thought it unsatisfactory to say that moral qualities are distinguished by sense or feeling rather than by reason. For after all it is a contingent matter what will produce certain feelings in us. God might have annexed colour sensations to the kind of light reaching our eyes in a quite different way from that which he has done, so that the very same things (in their real nature) which are red to us might have been blue, and vice versa, or we might have experienced a quite different range of colours in the same physical situations. If moral qualities have the same status as colours, it follows that God might have made hatred morally good, and benevolence morally bad, by attaching different sensations of the moral sense to them. There would have been no sense in which men with such a moral sense would have been mistaken, would not have known the real moral truth. But benevolence, according to these rationalist moral philosophers, is good in a much more ultimate way than this. So also are various other qualities of mind and principles of action, for, in the opinion of these thinkers, Hutcheson over-simplified and distorted the truth by treating benevolence as the one moral desideratum. That which is good or bad morally is so with a necessity which is apparent to reason just as are fundamental truths of mathematics. These things could not possibly have been otherwise, and even God is subject to these necessities. God's own moral sentiments are correct because, as the perfect being, he has perfect insight into what is independently morally good and bad, right and wrong. The most brilliant proponent of this rationalist position was Richard Price (1723–91), whose viewpoint is somewhat similar to that of W.D. Ross.

The contrast between the view that moral distinctions are detected by sense and feeling and the view that they are revealed as necessary truth to reason was a central theme in the moral philosophy of David Hume (1711–76), whose philosophy is recognized as empiricism brought more or less to perfection. Hume held that moral distinctions must be revealed to sense, feeling or emotion, not reason, because only emotion can prompt to action as evidently moral thinking does. Reason, whether as insight into necessary relations between ideas, as in mathematics, or in the form of inductively based prediction of consequences, simply reveals the means to achieving what emotionally one wants to achieve. Moreover, Hume famously insisted, there is a prima facie gap between any assertion as to what is the case and any assertion as to what ought to be done, and this is best understood as the gap between reason's detection of how things are and passion's emotional response to their being so. Ignoring the gap between *is* and *ought* is committing very much what we have seen Moore and other intuitionists describing as the naturalistic fallacy. However, Hume's way of explaining the gap was not that of the intuitionists, but rather that of the emotivists. For the strongest plank of the attitude theory of ethics lies precisely in the claim that if ethical propositions merely report some special kind of fact, then the special relation between moral judgement and action is inexplicable. Hume's complex moral philosophy cannot simply be equated with emotivism, but it has much in common with it. A main difference is that Hume, like Hutcheson, was confident that the basic moral attitudes or feelings of men would be uniform once they agreed on the facts of a situation. So although in a sense one's moral judgements indicate simply one's personal emotional response to a situation, in making one's response stable by further enquiry into the details of a situation, and by further reflection, one is not only stabilising one's own feelings but moving towards the way of feeling about the situation on which men in general would converge. (Hume adumbrates here an idea developed later by Adam Smith (1723–90) that the moral qualities of action are a matter of how an ideal detached but sympathetic observer would feel about them.) Thus Hume does not sharply distinguish between interpreting moral judgements as expressing one's own feelings, stating one's own feelings, and suggesting how men in general would feel about a situation if they knew enough.

Hume gives an elaborate account of the psychology of moral

feeling. Virtues are essentially qualities of character which are useful or agreeable either to the agent or others. That we react with positive moral emotion to such qualities of character, even when they are not useful to us, turns on a mechanism of sympathy, or the tendency to fall into the feelings of others. In this connection Hume propounds a kind of utilitarianism for which the good is essentially the useful, in terms of promoting human happiness. However, to talk of its goodness is not to talk merely of its utility but to report upon or express or predict emotional responses thereto. Hume's moral philosophy is mainly a development, on an effectively non-theistic basis, of Hutcheson's and Hutcheson had already advocated a kind of utilitarianism, for in identifying moral goodness with benevolence, he had seen the goodness of a man as essentially the amount of happiness he produced divided by his opportunities. However, Hume thought that qualities other than benevolence were virtuous if they produced the feeling of approval. Such qualities include not only those which are naturally useful or agreeable but also what he described as the artificial virtues, in particular justice. These are habits of conformity to rules which have developed by a kind of convention or tacit understanding, based on general recognition of the utility of establishing and retaining them.

The main difference between the utilitarianism of Hutcheson and Hume, and that of Bentham, is that for the former it figured mainly as an account of our actual moral feelings, whereas for the latter it was not only this but even more the basis of a critique of current morality.

Joseph Butler

An especially impressive moral philosophy of this period was that developed by Joseph Butler (1692–1752), later Bishop of Durham, in a series of sermons. Butler gave what have often been considered decisive arguments against the doctrine of psychological hedonism. He contended that pleasure could only arise from the satisfaction of an impulse and that therefore there could be no pleasure, and no possibility of satisfying a desire for pleasure, unless there were impulses towards things other than pleasure. Thus we get pleasure from such different things as food and fame, because we have a direct desire for these things.

Besides the particular passions or impulses directed at objects other than pleasure, a human being has a more general desire that he should enjoy as much pleasure or happiness as possible, during his life as a whole, which is as much as to say, a desire that the totality of his impulses directed at objects other than his own pleasure should receive as much satisfaction as possible. This more general desire is called self love or prudence.

It once being realized that one can desire things other than pleasure, there is no reason to fudge the fact that people evidently have, to some extent, a desire for the happiness of others, whether of people in general or of particular people. This is called benevolence, more especially so when it takes the broad form of a wish for the happiness of others in general.

Basic impulses towards such things as food, drink, warmth and sex are called particular passions by Butler, and stand in contrast to self love (or prudence) and benevolence which are the desire to have a maximal balanced satisfaction of the particular passions either in ourselves or in others. (However, benevolence is sometimes classified rather as a particular passion.) Butler insists that a particular passion, like that for some particular food, say an apple pie, is not directed at self at all, but at some external object, such as the pie. This seems a little misleading, for one's impulse finds satisfaction not merely in the existence of the pie but in the eating of it. Butler sometimes talks of a passion as directed at one's 'having' its object, but it seems more satisfactory to say that particular passions are directed at occurrences, and that the basic one cannot be directed at the occurrence of pleasure for oneself since this always stems from having something occur one wished to do so. He can then argue that once one has advanced beyond the superstition that all desire is for oneself having pleasure, one has no reason of principle for denying the apparent fact that some desire is directed at occurrences not involving oneself at all, but rather the welfare of others.

Butler's famous refutation of egoism is hardly conclusive. For one thing he certainly over-states it when he says that pleasure always comes from satisfaction of an antecedent desire for something other than pleasure. A baby surely finds pleasure in being tickled without having first desired it, and we may similarly enjoy a sudden glow of warm sunshine on our back which we had not sought. And the view should not be ruled out too summarily that all our desires grow up from the fact that certain things have been found immediately pleasurable. Moreover, even if particular impulses are not mainly directed at pleasure, the typical ones do seem to be directed at the occurrence of certain states of ourselves, so that even if not hedonistic they are egoistic. Besides which, Butler's argument really moves at the level of phenomenology only, as an account of the conscious character of desire, and hardly takes on the idea of someone like Spinoza that all activity at a deeper level is a manifestation of the organism's disposition to preserve and enhance its own being. Nonetheless, Butler showed that there is little justification for taking egoistic hedonism, or even an egoistic psychology which is not hedonistic, as based on obvious fact, and that the initial presumption must be that concern for the welfare of others is what it seems, the direct wish for their welfare.

In his more positive ethical viewpoint Butler tries to give sense to the old opinion that one should live according to nature. This is obviously a wrong maxim, he points out, if it simply commends doing what one is most inclined to do. However, it can be understood as recommending that one live in such a way that the various aspects of one's being are given the same degree of power as they have of innate authority. He points out that even the proponent of a purely egoistic view of life effectively acknowledges that different aspects of our being have different degrees of authority. We think it against nature if someone so lacks prudence that they involve themselves in great foreseen evil for the sake of satisfying some fairly trifling present impulse. Even the egoist sees that it is unnatural (however common) to let a particular desire weigh against the judgement of self love as to one's best long term welfare. Thus he recognizes that it is built into our nature that self love has authority, and therefore should have control, over particular impulses.

In the light of that recognition, he cannot dismiss as mere fantasy the sense we all have that it would be an unnatural organization of our nature if some slight personal desire was given its head against benevolent concern with some major aspect of the welfare of others. Hume was to say that it was not against reason to prefer the destruction of the world to the least uneasiness in one's little finger; it would simply be rather bizarre and repellent to normal feeling. For Butler, however, it would defy relations of relative authority, between our different psychological components, which are essential to our human nature.

Thus we have particular impulses and two higher level organizing principles of self love and benevolence. It is often quite appropriate to act on particular impulses but it is against nature to let these have their head against the opposition of the higher principles, or not to let these higher impulses remain upon the lookout for trouble. But what if the higher order impulses conflict? Butler believes that there is a further principle of our nature, conscience, which has a still higher degree of authority, and which will judge between all the lower levels of our being, including self love and benevolence.

Butler feels confident that, although self love, benevolence and conscience are distinct components of our nature, properly operative on different occasions, they would, when properly enlightened as to facts, give the same directives. This is partly a matter of religious faith concerning the world to come. However, Butler argues (rather effectively) that the belief that they are often in conflict, even in this world, is a mistake. He stresses in particular that enlightened self love itself will realize that too much love of self, and too little of others, stands in the way of happiness.

A question not very satisfactorily treated is as to what aspect of our being recognizes that there are these different degrees of authority, and what sort of reality *authority* is. An emotivist will think that Butler is merely expressing and inviting an attitude whereby we favour this sort of system of self control, while an intuitionist will think that there is tacit appeal to an intuition as to what ought to be. Perhaps Butler's position is essentially that of such an intuitionist. But one can still ask what part of our being it is which makes the judgement as to where the authority lies. Butler's view is that it is conscience, which he also equates with rationality, understood both as the faculty of discerning what is right, and the disposition to act in accordance therewith. However, it hardly seems satisfactory to say that it is conscience which tells us that conscience should be at the controls, for presumably self love would say the same of itself if given its head.

It might have been better for Butler to have set *reason* apart from the whole hierarchy of conscience, self love, benevolence, and particular passions and allotted it the role of ascribing different degrees of authority to each, that is, of saying how they ought to relate to each other mutually. Then conscience would not so much be a faculty of intuiting truths as a God-given power within us which – if put in control – will steer us in a particular direction, when it comes to choosing between ourselves and others. Another approach would be to say that it is *I*, not some *part* of me called 'reason', which sees that this is how the system of my impulses should be organized. However, this is on the way to casting doubt on Butler's whole tendency to divide the person, along lines rather similar to Plato's, into different faculties.

Butler's subtle moral philosophy represents the quiet wisdom of the Anglican church at its best, and has much to commend it even when detached from any doctrinal Christianity. The sermons in which it is presented are arguably the finest contribution to moral philosophy in English.

Kant

Kant's moral philosophy is sharply opposed to the moral sense approach of Hutcheson and Hume. It has something in common with the rationalism of Richard Price, but provides a kind of general formula for testing the morality of actions as Price (and later Ross) did not. Hume and Kant (1724–1804) are two towering giants of philosophy and our brief indications of their positions on morality – which must tear them from their systems as a whole – may irritate the knowledgeable, but with the space at my disposal and the particular purpose of supplying background to Part Two of this work, the only alternative was to ignore them altogether.

In our brief consideration of Kant's moral philosophy, it may be helpful to begin by distinguishing three different sorts of questions – of which the following are chief examples.

(1) Are we under any moral obligations – or, otherwise put, are we subject to a binding moral law? How can we know whether we are or not?

(2) Supposing that there is such a moral law and that we are

under moral obligations, what is their content? How can we determine what is and what is not obligatory? In short, what is the mark of the moral as opposed to the immoral?

(3) What is it for someone to act with a view to his moral obligations? What is it to act on the basis of moral, as opposed to non-moral, considerations?

The main interest of Kant's ethics lies in his answers to the second and third sort of question. However, while for many philosophers, especially in the present century, these are two different sorts of question, for Kant they are essentially one. For he does not allow that there can be someone who genuinely asks himself what he should do in a moral way, but whose answer is in terms of what is only one of various logically coherent moralities. Such questions are worth distinguishing, however, because there are moral philosophers today who think that Kant effectively answered the third correctly, but that he offers no sufficient answer to the second. A chief example of these moral philosophers is R.M. Hare.

Kant answers the first questions by contending that we cannot strictly speaking *know* that there is such a moral law. We can, however, have a reasonable faith that there is, based on the realization that without it all moral thought is a complete illusion. Doubts as to even the possible reality of such a law, arising from an excessively empiricist conception of the possibilities of being, prove unreasonable in the light of the establishable fact that both the every day world in which we live, and we ourselves, are only appearances of a realm of things in themselves whose true nature is hidden from us. For this opens the possibility that what we are in ourselves is essentially rational beings, belonging to a society of rational beings, while what we are as appearances is sensory beings. In view of that, we can recognize the possibility that as rational beings we fall under a system of law which we have somehow ourselves brought into being, and that it is our task while appearing to exist in the sensory world to live according to that law, in spite of the fact that what we appear to be is simply animals driven by sensory desire. The possibility that we belong to such a so-called noumenal realm (that is, a realm of things in themselves) in our true being, also suggests dimly how it can be possible that ultimately we are free agents, who can cause ourselves to act according to the moral law, whatever the pressures upon us, in spite of the fact that at the level of appearance we are simply parts of the natural order of cause and effect, and as such merely animals impelled by our instinctive desires.

Supposing then that there is such a moral law, what is its content? The first thing to realize is that it must consist in *categorical* imperatives, which are to be distinguished from *hypothetical* imperatives. A hypothetical imperative tells me that, as a rational being, I must do such and such if I desire to obtain a certain upshot, because it is the essential means to that upshot, and he who wills the end must will the means. Thus the doctor who tells me to take a drug, if I wish to recover from a sickness, is expressing a hypothetical imperative. In contrast, a categorical imperative bids me, as a rational being, do such and such, whatever my desires may happen to be, because doing so is essential to living up to my status as a rational being. As such, I will realize that I *must* do it, whether I wish to or not.

The first thing to realize (in Kant's opinion) about moral action is that it is action done out of respect for duty as such, that is for a system of categorical imperatives recognised as binding on every rational agent whatever his desires and feelings happen to be. In this connection Kant is somewhat notorious for sometimes seeming to think it better to help those in need out of a sheer sense of duty, without feeling any sympathy for them, than with feelings of loving kindness. He certainly holds that as long as my action is prompted by mere feeling, or by a judgement as to what will best satisfy my desires, I am not in sight of moral living. Even if such feelings and judgements are not disreputable in themselves, they must be under the firm guidance of respect for sheer duty. Most philosophical systems of ethics, and most popular moralizing, are radically flawed because they recommend morality to us either as what it is in our own best ultimate interests to do, or alternatively try to promote it by appeal to our feelings, for example feelings of compassion, or (like Hutcheson) by reference to some kind of moral sentiment which just happens to be part of human nature. These theories make morality *heteronomous*, that is, a system of principles whose validation lies outside itself, as opposed to *autonomous*, that is, valid of its own nature.

But how can I know what the categorical imperatives binding

on me are? The answer is that in a certain sense there is only one categorical imperative, and all other categorical imperatives are applications of this. This basic categorical imperative bids me only act on such a maxim as I can will should become a universal law to which all rational agents conform their behaviour. The maxim on which I act at any moment is the personal rule which is guiding my behaviour – for Kant it is of the essence of voluntary action that there is such a maxim in every case. The verbal expression of a maxim will be something like: 'In circumstances of such and such a sort I shall act thus', and I act immorally unless I can will that all agents should guide their behaviour by a corresponding maxim directed at themselves. That the categorical imperative bids me act in a particular way follows from the impossibility of avoiding the action without acting on a maxim I cannot thus will universalised. That I cannot may have either of two causes. It may be that the state of affairs in which everyone (as opposed to just some people) is guided by such a maxim is an impossibility, or it may be that it is in some way a thing that no individual can wish for. Kant thinks, for example, that one who breaks a promise, because it is going to land him in personal difficulties to keep it, cannot will that everyone would break their promises in these circumstances, for the situation in which no one kept promises which it turned out in the least difficult or vexatious to keep is an impossibility. This is because there could be no institution of promising in a society in which this was the deliberate practice. An example of the other kind is the obligation to help those in need. It would be possible to have a society in which it was everyone's personal rule to refuse all help to the needy, but as someone who might be needy myself I cannot truly will that all should turn their back upon me in such a situation.

Although Kant has little respect for most previous moral philosophies, he has immense respect for the untutored moral insight (if not practice) of the ordinary man. In spite of all that has been said by popular moralists, along the lines of honesty being the best policy, everyone really knows implicitly that it is by this test of universalisability that one should determine what one ought to do. It is also how one must determine whether others have acted well or not, so far as externals go, though how far they have acted well in a proper inward sense, that is, how far they have been truly guided by the categorical imperative, rather than by the calculations of self interest, is hidden away in the depths of their being, hidden perhaps even from themselves. Thus Kant's moral philosophy has the merit of mainly being concerned with illuminating how each of us should decide what to do himself, rather than with advising him how to judge others.

Kant gives several alternative formulations of the basic principle of the categorical imperative, which he believes have the same essential content but present different aspects of it to our thought. The other main formulation is the rule of acting in such a way that you treat humanity, whether in your own person or in the person of any other, never simply as a means, but always at the same time as an end. That every person is an end in himself or herself is related to the fact that the only thing which is good without qualification and in all circumstances is a good will, something which everyone has the potential to *be*, whatever talents or gifts of fortune they may or may not possess. It is in virtue of this potential that people matter in a way in which no *thing*, or mere *animal*, can do.

This hardly looks the same principle, but the connection lies in the fact that for Kant the sense in which every person is an end is that each is a rational agent who, as such, should be conceived as potentially cooperating with me in settling upon and living by universal principles of behaviour taken as binding on all rational agents. (Indeed at the level of ultimate reality we have perhaps already agreed upon such a system as our guide within the phenomenal realm.) Thus if I break a promise for my own convenience, I fail to treat the person to whom I made it as an end in himself, for I can hardly expect him to endorse a principle of action which allows him to be treated thus.

Many feel a certain warmth for Kant's idea that the only thing which is good without qualification and in all circumstances is a good will, without accepting that a good will is one concerned only with the performance of duty for duty's sake. They may feel that a good will can take the form of seeking to relieve suffering out of sheer compassion, without thought of duty, or of a direction of intent upon 'the good, the true, and the beautiful' in whatever concrete shapes they present themselves as realizable.

Kant's approach to ethics is like utilitarianism (at least in certain versions) in just one way. It professes to offer a rational decision procedure for testing the morality of actions, possessed of some kind of objective status, which means that it is irrational to reject it. It is, however, most unlike utilitarianism in almost all other respects. Indeed, utilitarianism and Kant's ethic of duty for duty's sake are usually seen as the two great extremes of moral theory, the line between them representing a continuum on which most moral philosophies with a positive theory of what is right and wrong can be placed. Utilitarianism is often thought of as challenging us to rethink the ethics of the common sense of our society. Kant does not think he is putting forward a doctrine on the basis of which ordinary morality can be criticised, but that he has formulated the principles by which all good people implicitly know they should, and in their hearts do, judge their conduct.

In spite of the unlikeness of Kant's approach and that of utilitarianism, one of the foremost moral philosophers of our time, R.M. Hare, has developed a moral philosophy which is something of a synthesis of them. In his earlier work Hare seemed to believe that there were alternative moral stances to the world, each of which might be equally rational, though we may quite properly condemn some of them from the perspective of the morality which we ourselves advocate. Hare then held that one was not, so to speak, in the arena of moral debate at all, unless one's moral judgements were based on considerations one was prepared to universalise (in a sense close to Kant's). However, different people might without illogicality be prepared to universalise quite different considerations, so that there could be alternative moral views, each equally a *moral* view, even if some of them were, from our point of view, immoral. But gradually, as I noted in Chapter One, he came to think that one cannot universalise judgements coherently, without this amounting to the development of a universal sympathy with the desires of all affected by one's action. For one has to recognize that if one had their desires one would not accept principles which rode roughshod over their satisfaction, and this implies that one should not accept them at all, since one cannot universalise them to that hypothetical situation in which one would be forced to reject them. So one must take into account the effects of one's actions on the satisfaction or dissatisfaction of the desires of all those affected, along the lines of preference utilitarianism.

When Hare speaks of universalisability he refers to the

requirement that one be prepared to support one's ethical judgements by precepts couched in completely general terms. This serves somewhat to clarify a point not made explicitly by Kant, but clearly intended, namely that in universalising one's maxim, one must purge it not only of the word 'I' understood as referring to one's own particular self, but of proper names, and so forth, referring to one's particular acquaintances. If I act on the maxim that I will not do things which will hurt Mary's feelings, that is not universalised simply because I will that all should avoid hurting Mary's feelings. To be universalised it must be transformed into some such principle as that no one should hurt the feelings of another person, in such and such generally describable circumstances.

However, many philosophers today would probably go along more with Hare's original position and say that although Kant may well have answered the third sort of question (page 98) well, he has not adequately answered the second (still less, it would generally be thought, the first). The most he has done has shown what it is to approach things from a moral point of view, he has not shown that only one particular sort of maxim can coherently be universalised in each case.

One of the earliest objections to Kant lies in his combining an objective view of ethics with the view that morality is properly expressed in imperatives. It is not as though his position is remotely like that of modern emotivists who have compared ethical statements with imperatives (and Hare's ethics differs most from Kant's where it comes nearest to emotivism). For Kant there is an objective moral order. But how can imperatives function as anything other than instructions directed at others (perhaps also himself) by an individual with status as a commander? Perhaps some sense can be made of regarding hypothetical imperatives as objectively true, since they can be taken as statements about what needs to be done to achieve something, but how can there be a categorical imperative, with no one to command it? True, Kant thinks that if morality is ultimately valid it is because we have somehow settled on these imperatives ourselves at the noumenal level, but to this – apart from the dubiousness of the metaphysics – it is likely to be objected that if the imperatives spring from myself it is quite proper for me to rescind them when convenient.

Schopenhauer

Arthur Schopenhauer (German philosopher, 1788–1860) developed a metaphysical account of the world which started out from the fundamental claim of Kant, that the world of daily experience is a world which only exists as an appearance presented to the human mind. It is somehow constructed by our minds, on principles which are innate to them, when those minds in their true being respond to the reality in the midst of which they exist in the true being of that. We have seen that Kant thought it a proper object of ethical faith that in our true being we are free agents who have somehow cooperatively bound ourselves to live, even in this realm of appearances, by a universal moral law. Schopenhauer largely rejected Kant's own dark suggestions as to what things in themselves might be like. He argued that there is but one thing in itself, a cosmic will which gives itself the illusion of living as a plurality of beings in a world of space and time. Each of us is this cosmic will imagining itself to be just one individual, sharply distinct from every other individual. This illusory pluralisation of itself is seen as a kind of fall of the cosmic will due to its hopeless craving for an impossible satisfaction. In man, at his most saintly, the cosmic will becomes conscious of the illusory and unsatisfying nature of plurality and moves towards a kind of reconstitution of its original and blissful unity before it began to crave. (Temporal expressions such as 'before' cannot be taken literally, since all this somehow occurs outside time.) Moral goodness is a phase in the development towards such saintliness in which we sense, and act on, our ultimate identity with all other conscious, and indeed unconscious, beings.

Strange as this view of things may seem, it is argued for by Schopenhauer with great forcefulness. However, our concern is with his more specifically ethical views.

Schopenhauer attacks Kant's conception of ethics on several grounds. He holds that the very idea of a categorical, as opposed to a hypothetical, imperative is an absurdity. An intelligible imperative is normally an order given by someone who can impose sanctions on those who do not conform to it, and has the form of 'Do this . . ., or else' and is hypothetical inasmuch as it tells one to do or avoid something if one wishes to avoid the penalty. Schopenhauer believes that the categorical imperative only seemed to make sense to Kant, because unconsciously he took it as the command of God. Besides, if one somehow legislates the imperative for oneself, as Kant says one does, why cannot one let oneself off the hook by repealing it?

Apart from this, Schopenhauer thinks that Kant is, in spite of himself, really giving ethics an egoistic foundation. For if one asks why, in the kinds of case Kant uses as examples, one is not prepared to see the maxim of a bad action universalised, one finds it is because one is, for egoistic reasons, not prepared to have the maxim rebound on oneself. Thus one is unwilling to see the maxims of breaking a promise when convenient, or of ignoring the sufferings of the needy, universalised because one envisages this leading others to breaking promises made to oneself or refusing to assist one when in need.

Schopenhauer seems partly right and partly wrong here. He is right that the reason Kant gives in his most persuasive examples why one would not be prepared to will the maxim universalised seems to be that it might then rebound on oneself. But that does not show that one who refrains from an act because for egoistic reasons he would not like it universalised is acting for solely egoistic reasons. After all, the sheer egoist might well act on a maxim universalisation of which might – in some circumstances – be unwelcome to him, either because he was confident those circumstances would not arise, or because he has reason to think that his living by that maxim will not encourage others to do so. This seems a weakness in Schopenhauer's criticism. But Schopenhauer's idea seems to be that even if this is so at a formal level, the real appeal of Kant's reasoning lies in its arousing our sense of the precedent that immoral conduct sets for maxims of action which others may use against ourselves. Even if this is somewhat unfair to Kant, Schopenhauer may still be right that egoism plays an uncomfortably large place in his ethical reasoning.

Schopenhauer's most effective weapon against Kant lies in the scorn he, like others, pours on his conception of moral goodness. The idea that the good man may very well have no sympathy for the sufferings of others, provided his eyes are fixed on the stern call of the categorical imperative, clashes with what Schopenhauer thinks the evident fact that humanity only ever ascribes moral worth to actions of which the driving force is compassion. For this reason Schopenhauer also attacks all ethical theories, which base ethics on egoistic considerations, even ones as lofty as Spinoza's.

To support this, Schopenhauer urges first that the following heads supply an exhaustive classification of human motivation: (1) egoism; (2) malice; (3) compassion. He argues for this by saying that nothing can in the end attract the will but the weal and woe of some conscious being, and that this may either be the weal and woe of oneself or another, and either be such that weal and the removal of woe are the goal or vice versa. For egoism and compassion it is the weal and removal of woe that attract, the difference being as to whether it is one's own or another's; for malice the goal is another's weal and woe, but with so to speak the positive and negative signs attaching to them reversed, a reversal which, in the nature of the case, cannot occur at the egoistic level. (Presumably Schopenhauer does not mean to deny that some actions stem from a mixture of these motivations.)

Having classified motivation thus, Schopenhauer asks us how knowledge of the kind of motivation present affects our moral judgement of an action. He claims that we all see an action as bad as soon as we know it is motivated by malice, and as morally indifferent once we know it is motivated by egoism, that being simply the normal, neither good nor bad, pattern of human life. It is only when we believe that an action is inspired by genuine compassion that we find moral worth in it. (It seems to me that Schopenhauer should add, as perhaps he really meant to do, that some actions are positively bad, not so much because they exhibit malice as because they exhibit an unusual lack of compassion.)

Schopenhauer's classification of motives may be over simple. However, if we take it as the claim that the positive or negative goal of action is always something conceived as a good or an evil, and that this good or evil must be envisaged as something either to be realized in one's own consciousness or in that of another, it is persuasive. Many, in particular, will agree with Schopenhauer that the notion of a respect for the moral law which cannot be explicated in terms of motivations falling under one or more of these three classes is empty, and will sympathise with Schopenhauer's claim that it is really either a form of egoism, or a special sort of compassion, not something standing quite apart from other sorts of motivation.

An objector might say that in allowing the existence of a malice for which the faring ill of another is a direct goal of the will, Schopenhauer has departed so far from the idea that all desire is desire for the good or the avoidance of evil, that he has given up the main ground there is for denying that there can be goals not concerned with the unhappiness or happiness of sentient beings. I have some sympathy with this point, but must hold my own views about motivation in reserve for now.

Schopenhauer goes on to develop an account of ethics which is simply an endorsement and elaboration of what he takes to be the universal human sense that moral worth consists in the extent to which our actions are inspired by compassion. He suggests that the concern of compassion is embodied in a precept which is the basis of all ethics, 'Injure no one; on the contrary, help everyone as much as you can.' (In the Latin which Schopenhauer likes to use: *Neminem laede; imo omnes, quantum potes, juva.*) He attempts to derive the whole of ethics from this simple foundation.

An immediate objection is that he has after all formulated the basis of ethics in that imperative form to which he took such exception in Kant. Schopenhauer would retort, I think, that although the grammatical form is imperative, it is really just a description of the motivation of actions possessing moral worth. For Schopenhauer it is simply a fact that some men are good, and others bad, in that some are more disposed to do acts possessing moral worth and others to do wicked acts. Part of what is involved in being good is that one will be concerned to find out what in detail one must do in order not to harm others, or to help them in their woe. However, the moral worth of the action lies in the fact that this is what one is seriously seeking to do, not in the success one has in doing it. Thus in describing how the good man will go about deciding what to do, one is not telling anyone to make decisions in this way. This would, according to Schopenhauer, be useless because the fundamental moral character of a man is unalterable.

The negative clause of this precept, 'harm no one' is the basis of justice. The positive clause 'help everyone as much as you can' is the basis of loving kindness. These are the two cardinal virtues from which all others are derivative, while they in their turn are derivative from compassion. When someone does that which, so far as externals go, is just what the good man would have done, but is motivated either by egoism or malice, they are not exhibiting these virtues, and their actions have no moral worth.

Schopenhauer's attempt to show that justice is derivative from compassion is not very satisfactory. He seems to take the precept 'harm no one' as meaning 'do not cause suffering, and as one specially important case of this, do not interfere with anyone's obtaining by their activity what they would otherwise legitimately achieve by it'. By an illegitimate achievement is meant one made at the expense of someone else's achieving what they would otherwise – judged by the same criterion – legitimately achieve by their action. The idea is that justice consists in one individual will not interfering with the pursuits of another individual will insofar as that will is not interfering with other wills except to stop them interfering with other wills which are not interfering with other wills which are not . . . (and so on).

Concerning this one may ask first whether justice as Schopenhauer here describes it is, as he thinks, simply a special application of compassion, and secondly whether Schopenhauer's account of justice is adequate to our ordinary conception of it. Perhaps we can grant him the first point, for the justice he describes is essentially the restraining of oneself from hurting others out of respect for the welfare of all. The answer to the second question is more problematic. It seems doubtful whether most people's conception of justice simply boils down to the desire not to hurt others. Schopenhauer's approach can be suggested by considering the justification he offers of self defence. He argues that if another tries to injure you, and you have to injure him to save yourself, this is really a case of the other person injuring himself, since the cause of a cause is the cause of an effect, and it was the attacker who caused your defensive action. Thus he has been hurt by himself not by you and you have done him no injustice.

Surely this is shaky reasoning. Even granted that the attacker caused your action it was still your action, as much as any other, and the question whether it was morally legitimate for you to respond cannot be resolved by the pretence that you are not a moral agent in the situation at all.

However, the defects of Schopenhauer's treatment of justice do not necessarily falsify his claim that the only ultimate merit exhibited in just action lies in the concern it shows for others, and that the same is true of all other forms of moral virtue. It is important to realize, however, that Schopenhauer is not professing to derive the institutions of the state, and the legal sanctions which support them, from men's compassion. Rather does he regard the state, in Hobbeist fashion, 'as the masterpiece of the rational and accumulated egoism of all which, understanding itself, has put the protection of everyone's rights into the hands of a power infinitely superior to that of the individual, and compels him to respect the rights of everyone else' (p. 129). Thus we have good egoistic reasons for maintaining the state and abiding by a system of rules which we all have good egoistic reasons for enforcing against each other. There is, however, no moral worth in all this unless someone sticks by what is called justice out of compassion for others.

The compassion which shows itself in justice, so far as that is a genuine virtue, is the disinclination ourselves to bring sorrow on others. So far as a man's behaviour manifests the absence of this disinclination, he behaves *wrongly*. Action cannot, however, be called wrong merely because it shows the lack of that more positive sort of compassion which actively seeks to remove sorrow of which the agent is not himself the cause, for such action is inevitably the norm. However, the highest sort of goodness is shown in a man who possesses also this more positive form of compassion, which constitutes the virtue of *loving kindness*. Schopenhauer notes that this virtue was not listed among the virtues by Plato or Aristotle or other ancient philosophers. The proper articulation of its significance only entered Western thought with Christianity, although, in the East, Hinduism and Buddhism had all along recognized its fundamental nature. But though the ancient philosophers did not list it, Schopenhauer tries to show by various quotations from ancient literature that it was given due admiration among the people at large. He claims that all the world over nothing is looked up to with greater admiration than the genuine concern of one man to alleviate the sufferings of others.

So Schopenhauer's main claim is that moral worth consists quite simply in compassion, which in its lesser form, as justice, stops one positively harming another, but in its fuller form, as loving kindness, carries over to an active concern to help others in their need. It should be noted that for Schopenhauer the goal of compassion is the relief of misery and does not include the creation of positive happiness. This is partly because his pessimistic view of life implies that positive happiness, as opposed to relief from the worst sorts of unhappiness, is impossible, and partly because he thinks that the kind of identification with others which constitutes compassion can only occur when one becomes aware of another as a sufferer.

Schopenhauer is a very great philosopher whose deep insights are continually marred by his misanthropic temperament. If we are to continue to learn from him, we must free the gold from the dross. Thus in evaluating his attempt to base ethics on compassion it is best to detach it from his intense pessimism about the possibility of positive happiness. In any case, the distinction between relieving sorrow and promoting happiness is hardly sharp, if it is true, as I think it perhaps is, that once a person's sorrow is removed (other than by death) what takes its place is necessarily happiness, while when happiness is absent sorrow is necessarily present. This being allowed, we may agree that the main task of moral goodness is to avoid causing, but rather positively to relieve, sorrow.

It may be going too far to say that a favourite doctrine of Schopenhauer's, that of the fundamental unalterability of moral character, belongs to the dross of his thought, for his treatment of the subject fits rather beautifully into his overall metaphysics of the will. (It may also stem partly from his sense of his own failure to attain the virtues his ethics celebrates.) However, there is no need to treat it as an essential part of the main claim that ethics rests upon compassion; indeed, that claim can be put to more effective ethical use without it. Still, a brief glance at this doctrine may be of interest.

It is related to his views on the question of free will and responsibility, views which are his own special development of Kant's position. Each action is causally determined in that it flows necessarily from the combination of the agent's character and his beliefs about the consequences of acting in one way rather than another. The beliefs are the cause of the action, but like all other causes, they only operate because they affect something with a determinate nature. At the level of causation specified in physics it is only the determinate nature of matter in general that is involved, while chemistry and biology explore the type of causation which arises in matter which has reached a higher level of complexity. The causation of human activity is just as inevitable but there is no one single set of causal laws, because each single human has a quite unique determinate nature. This is his moral character, the special quality of his will. It is this which is the ultimate possessor or otherwise of moral worth. A man is blamed not so much for what he does, but for what his action shows that he is. This cannot change, because all change in a man's outward behaviour arises from causes which can only operate on him in conseqence of his unchanging basic character. Causes only affect him as they do in virtue of his character and therefore cannot act on his character.

Ingenious as Schopenhauer's development of this theme is, it hardly seems finally convincing. It depends on treating each man's character as possessing a status like a basic force of nature such as gravity. Gravity cannot change (it may plausibly be said) because it is part of the basic background nature of things which settles the principles on which all change comes about. Schopenhauer thinks that the same is true of every individual human personality. But it is far from obvious that this is the best interpretation of what human character is, or one which fits well with developments in psychology and related fields since Schopenhauer's day. Moreover, in the end Schopenhauer himself attaches great importance to the possibility of one quite fundamental change, in which a man moves from an assertive mode of being in the world to one in which the principle of his life is renouncement. So an ethic which is essentially Schopenhauerian can, whatever Schopenhauer says himself, allow that moral worth may develop or atrophy.

If that is so, someone who accepted an ethics of compassion could look again at the role of the capacity of ethical thought to influence human life for the better. Schopenhauer himself allows that

> through an increase in insight, through instruction concerning the circumstances of life, and thus by enlightening the mind, even goodness of character can be brought to a more logical

> and complete expression of its true nature. This happens, for example, when the remoter consequences that our action has for others are pointed out to us, such as the sufferings which come to them indirectly and only in the course of time, and which arise from this or that action that we did not consider to be so bad. (p. 194)

Surely the acquisition of a sense for the effects of his action on others can actually develop the capacity to feel and be influenced by compassion, and thus to become a better person. Schopenhauer's insistence that the capacity is a *constant* representing a person's degree of innate unalterable goodness, which is simply called forth into more effective action by developing knowledge, seems a rather specious device for protecting the doctrine of the unalterability of character. If so, an ethics of compassion may see moral education, and ethical reflection, as directed at the development of this capacity for compassion by bringing facts about the world before people in a sufficiently vivid way.

In doing so, it would be providing a rational foundation for ethical recommendations, insofar as these would be based upon rationally established facts about the world. But would this not be simply a matter of the contingent fact that in some people such considerations cause the moral attitude of compassion? Would not the malicious person in whom awareness of the same facts simply served to point out further opportunities for being cruel be just as rational as the compassionate person who was prompted to acts of charity?

Schopenhauer's metaphysics gives a negative answer. There is a deep truth about reality of which the egoistic and still more the malicious are ignorant, namely that the apparent distinctness of oneself from others is an illusion. The man who compassionates others implicitly senses that it is really the same self which suffers in another as is present in him, so that his cruelty or indifference is directed at himself. It is only because one thinks that the other is something fundamentally different from oneself that one can seek to hurt him or be indifferent to his sufferings. Recognition of the deep identity between us cannot but bring the same kind of concern for others as one has for oneself. Thus wickedness is a kind of ignorance.

Schopenhauer would not put this ignorance down to irrationality in any usual sense, and would not think that it could be put right by reasoning. The deep truth is one revealed to the good man's heart rather than to the metaphysician's intellect. But even if this is so, the metaphysician – if we accept Schopenhauer's metaphysics – is rationally demonstrating a truth in virtue of which the good man can be seen to possess a better understanding of how things are than the bad man. So in a sense Schopenhauer has found a rational foundation for ethics, to whatever extent his metaphysics is correct.

What may well be doubted is whether he is much help in the quest for a decision method for resolving difficult ethical questions. How would his views bear, for example, on such a modern issue as to whether experimentation on embryos is ever legitimate, or whether surrogate motherhood for payment should be allowed?

Presumably the good man, as Schopenhauer conceives him, would decide such questions in much the utilitarian's way, since both give pride of place to doing what will minimise suffering. However, Schopenhauer is concerned to describe the kind of motivation which has moral worth, not to tell someone thus motivated what to do. But if we do try to extract directives from Schopenhauer, what we get is the commendation of a compassionate approach to life, in which we respect the drive of each will for satisfaction and away from pain. Perhaps Schopenhauer would have said that an embryo or a foetus becomes an object of moral concern once there is ground for believing that a unified will to live pertains to it (which, for Schopenhauer, would imply that its behaviour is no longer wholly explicable by the general laws of nature) but clearly this raises and leaves many problems. Still, even if Schopenhauer's ethics, like the Christian injunction 'love thy neighbour as oneself', does not provide a complete guide to conduct in all circumstances, it is far from lacking all definite content.

Some affinity between Schopenhauer's views and those of utilitarians is further evidenced by the fact that they are the two main historical ethical theories which have raised duties of kindness towards animals to a prominent place. Kant viewed animals as non-rational creatures who could not be ends in themselves, or thought of as fellow legislators of the moral law with us. The only moral reason for kindness towards them is that unkindness to them would, by contagion, promote unkindness to humans. For Schopenhauer animals are simply humbler individual wills, like us ultimately aspects of the being of the one cosmic will which manifests itself in this illusory world of plurality.

To conclude this account of Schopenhauer's approach to ethics I quote a passage which shows him at his most sensitive and impressive. One suspects that it reflects personal experience of his own better and worse side as many readers may well feel it does for them too.

> The *bad* man everywhere feels a thick partition between himself and everything outside him. The world to him is an *absolute non-ego* and his relation to it is primarily hostile; thus the keynote of his disposition is hatred, spitefulness, suspicion, envy, and delight at the sight of another's distress. The good character, on the other hand, lives in an external world that is homogeneous with his own true being. The others are not a non-ego for him, but an 'I once more'. His fundamental relation to everyone is, therefore, friendly; he feels himself intimately akin to all beings, takes an immediate interest in their weal and woe, and confidently assumes the same sympathy in them. The results of this are his deep inward peace and that confident, calm and contented mood by virtue of which everyone is happy when he is near at hand. When in trouble, the bad character has no confidence in the assistance of others; if he appeals for help, he does so without any assurance; when he obtains it, he accepts it without gratitude since he can hardly understand it except as the effect of the other people's folly. For he himself is still incapable of recognizing his own true nature in that of another. That is the real reason for the revolting nature of all ingratitude. This moral isolation in which the bad character essentially and inevitably finds himself can easily drive him to despair. The good character will appeal to the assistance of others with just as much assurance as the consciousness he has of his readiness to give them his help. For, as I have said, to the one man, humanity is the non-ego, but to the other, it is 'myself once more'. The magnanimous man who forgives his enemy and returns good for evil is sublime, and receives the highest

praise, because he still recognized his own true nature even where it was emphatically denied.

It is an inadequacy in Schopenhauer's ethics that it leaves the status of his attribution of moral worth to compassion and to nothing else rather obscure. Remembering our G.E. Moore we may ask whether it is just saying that compassion is compassion and different from egoism and malice. So far as he has a view on the matter it seems to be that the moral worth of compassion consists in the quite special feelings of approval we all have towards it. This approval is given greater significance by the fact that the man who tends to feel such compassion is the man who sees deepest into the ultimate nature of reality.

Among likely criticisms of Schopenhauer's views are these three:

(1) It may be said that it is a very limited view which can see no human virtue except in altruism.

(2) It may be said that we should include under the heading of ethics the ability to organize our life towards the goal of self fulfilment and that this should be taken as a goal of twin importance with that of helping others. After all, unless there is some good which each does well to seek to realize in his own life the whole affair of mutual aid has no ultimate point.

(3) It may be said that Schopenhauer contrasts egoism and altruism far too sharply and is wrong to think all attempts to recommend good behaviour on the grounds that it assists the agent to live a fulfilled life beyond the pale of morality.

These criticisms are, of course, closely related.

(1) An extreme form of the first criticism is found in Nietzsche, whose philosophy is both a development of and a reaction against Schopenhauer's. According to Nietzsche 'good' and 'bad' only refer to unselfishness and selfishness in a slave morality by which the weak try to make the strong ashamed of themselves and to motivate them to help the weak out of compassion. In sturdier ancient times, and perhaps in a better future time, 'good' referred rather to a virile capacity to live one's own life to the full. As long as those capable of making something positive of their lives are weakened by impulses of pity and shame at their own self assertiveness, the great human achievements which are the one thing which really give point to the otherwise sorry

spectacle of the human world, will be shamed out of existence. Thus we will live more and more in a molly coddling welfare state in which the capable will waste their powers in keeping some semblance of happiness going in feeble second raters who will never make anything worth while out of their lives in any case.

(2) A more sober objector might take his stand on the second criticism and allow that compassion possesses real moral worth, and even that it shows a sense that the boundaries between one person and another are not ultimately real, but insist nonetheless that we should not restrict the kind of high praise expressed in a word like 'moral worth' to compassion, but should recognize the positive glory of personal achievement, and that sometimes people do better to develop their own powers, or help others do so, rather than be constantly bogged down by pity for the less fortunate.

(3) There seems something in the idea that Schopenhauer's enormous contrast between egoistic motivation and altruistic motivation itself reflects an exaggerated notion of the contrast between one person and another. For it implies that there is a sharp distinction between aiming at one's own good and that of others. Is there not something in the idea of Spinoza that ultimately the good of one is so much intertwined with the good of others that egoistic motivation, when properly enlightened, passes of itself into concern with the good of others? This point of view is developed more completely in the notion of self realization found in the philosophy of Hegel, F.H. Bradley and other absolute idealists, who believe that our basic impulse is to realize ourselves, but that the self we realize is essentially a social self, whose good cannot be separated from that of those among whom we live. It stultifies ethics, and falsifies our relations with others, so it is argued, to put some absolute separation between one's own personal and inevitable wish for personal fulfilment and what one is bidden admire under the name of virtue. After all, even Schopenhauer himself, in passages like that just quoted, gives us a sense that the good man experiences the world in a more satisfactory way than do others.

I believe that there is a good deal in these criticisms, and that an adequate view of ethics would not take Schopenhauer neat, any more than it would take Spinoza neat, but would be better found in some kind of synthesis of these two great moralists who, in metaphysics, were not so far apart.

F.H. Bradley

I move now to a a brief discussion of the moral philosophy of F.H. Bradley (1846–1924). Bradley owed a very great deal to the figure who dominated nineteenth century philosophy, G.F.W. Hegel (1770–1831). However, as my own style of philosophising is to a very special degree influenced by Bradley I feel it appropriate that I should take Bradley, rather than Hegel, for discussion.

In his *Ethical Studies* Bradley takes his departure from the presumed fact that every individual is struggling for some kind of self realization, and that this is the basic driving force of the attempt to live a morally good life. Such a life is not a means to self realization, but one main form of self realization. By self realization Bradley seems to mean the giving some sort of over-all coherent pattern or structure to your life in which you can find satisfaction, and such that all the details of your life are enjoyed as particular elements in that total pattern. One could perhaps say that it is life in the light of an accurate self image with which one can be satisfied.

If this is how we should conceive the morally good life then many traditional theories characterise it quite inadequately. From the standpoint of utilitarianism someone has a *personally* good life if they enjoy as much pleasure as possible, and lead a *morally* good life if they maximise the pleasure of all affected by their action. But this (so Bradley argues) is a hopeless criterion, for the very idea of a maximisation of pleasure is meaningless. If one postulates free will, one could presumably have always got or produced more pleasure; if one does not, one gets and promotes the maximum possible (because the only possible) whatever one does. Above all, a life devoted merely to piling up pleasures and keeping pains at bay has no structure such as we look for in the good life. But the good life is hardly better characterised by the Kantian notion of duty for duty's sake. This presents a mere abstract form of rationality from which one can only draw any concrete content by a sleight of hand. It is said, for example, that the reason for not breaking promises is that it is impossible for everyone to do this, since the very possibility of making promises depends upon their usually being kept. But one might as well say that one ought not to relieve poverty because it is impossible for everyone to do this, since there would be no poverty to relieve in a society in which everyone was disposed to relieve it.

A more satisfactory notion of the morally good life is that which equates it with the playing of one's proper role in a community. In a satisfactory community, and only in such is a morally good life possible, each person has or eventually finds their own appropriate role, in terms of job, role as husband or wife, as parent, citizen, member of a political party, and so forth. The good life as a contributing member of a society with a definite role gives a structure to one's life in which one can find satisfaction, and one's duties are precisely what is required for the playing of one's role. Bradley described this view of the good life as that of 'one's station and its duties'. This has for some of us today a rather unattractive ring, suggestive of 'the rich man in his castle, the poor man at his gate'. However, the essential idea that human beings find their self realization as members of a community in which they are playing a definite and respected role, does not necessarily imply that these roles take forms turning on social inequalities. Today, when the *anomie* of the unemployed is one of our main social problems, we should find this point of view of considerable relevance.

An obvious criticism of the notion that self realization is to be found solely in a social role is that some societies impose social roles whose functions are positively immoral. Sometimes the moral need is to change society. And, as a separate point, there seem to be some legitimate ways of seeking self realization which are not attached to social role.

Both these points are granted by Bradley. The ethic of my station and its duties is on the way to an adequate ethics but is not the last word on it. One point which he might have made is that *the reformer* of one kind or another is a definite social role – or a definite form which certain social roles can take – so that to the extent that a society has moral worth it will always be in a process of gradual transformation for the better by the efforts of those who play the role of reformer within it. As for what makes a society good or not, it is essentially that it embodies a form of life providing roles for its members which allow them to give a pattern to their life in which they can find satisfaction and a sense of developed identity, that is, a sense of being something definite, and of having made their own definite contribution to the life of the society.

Bradley sees humanity moving towards a cosmopolitan morality in which people find themselves at home by playing roles in a society which is ultimately that of the human race. En route to such a society, one (as one might say) tribal society can be better than another, if it is that into which the latter must be transformed if contradictions present in its own morality are to be resolved (or if it would be reached by a series of such transformations). (This Hegelian idea does not have to be given the specific content it receives in Marxism, but it does show that Bradley's thought is not essentially conservative, as it is sometimes thought to be.) However, the moral worth of an individual person turns not on the goodness of the society to which he belongs but on the success with which he lives up to the best lights of his own society.

Yet Bradley stresses the point that some few individuals can only realize themselves fully at the cost of some lack of commitment to the playing of a purely social role. (This theme is developed especially in some of his later writings.) For example, some creative artists and thinkers may need to turn their backs on society. If they do achieve self realization of some sort in a solitary quest this is likely eventually to enrich the lives of others, but even if sometimes it does not, they may have been right to pursue their solitary path. Bradley's metaphysics assures them that in some way their efforts will not have been in vain but will form part of a divine life in which ultimately we are all elements. Bradley believes this on the basis of commitment to a monistic – in a sense pantheistic – metaphysics having something in common with Spinoza's.

There is much that is wise in Bradley's treatment of morality which cannot be captured in a brief summary. He does not have the kind of snappy formula offered as the key to ethics by some of the thinkers we have considered. He explores the psychology of the development of moral character in a penetrating manner. He thinks of the child as developing a good self which is the main integrated personality, and a bad self which is a chaos of desires which cannot be brought into that structure. For most of us our bad behaviour is what falls outside the main self image which guides our life. Those who do not have this kind of personality structure at all are not so much immoral as amoral.

Bradley's strong emphasis upon the importance of social role, and the similar emphasis of Hegel, from whom he largely derived it, has seemed unattractive to those who have found the existentialism of such as Jean-Paul Sartre appealing. Existentialism regards it as bad faith to take one's morality from one's society as a mere given fact. The morality by which one lives must be one's own free decision, whether it is the morality of one's society or not, and one should not hide one's radical freedom from oneself by treating a conventional morality as binding on one other than through one's own choice.

The existentialist call to be strong in a sense of one's own freedom has its value, but those who seek a rational foundation for ethics wish for guidance on how (at least in contexts which they regard as 'moral') to use their freedom – in any sense in which they possess it, a matter on which there is considerable disagreement – resting on some kind of justified insight into some truth about the nature of the human situation. For Bradley the relevant truth is that the self realization which we are all really after is best found, (for the most part) in playing a recognized social role within the larger life of humanity. For Spinoza it was the truth that a certain kind of virtuous life will best enable one to actualise one's rational essence, and will perhaps culminate in a deeply satisfying sense of oneness with the cosmos. For Schopenhauer it is the truth, of which the good man is cognizant in that immediate intuitive way required if it is to affect action, that fundamentally it is the same self as looks out from another's eyes as is present in himself.

Two American moralists

I will conclude with a brief notice of the approach to the foundations of ethics of two American philosophers belonging to what has been called the golden age of American philosophy.

Josiah Royce (1855–1916), in his *The Religious Aspect of Philosophy* poses the whole problem of ethics in a deft way. Moral philosophies, he tells us, divide into those which seem to rob ethics of all motivating force by tying it up too much with rational judgement and others which give it motivating force by basing it on will, but which thereby seem to make it essentially a matter of arbitrary choice. His carefully worked out resolution of this dilemma is that motivation must indeed come from the agent's own will, but that we can only steer ourselves to our heart's desire by knowledge of the world, and that this knowledge informs us of the willing of others. To whatever extent we really learn the facts about another's will, and this matter of fact really sinks into our consciousness, we cannot but adopt the satisfaction of that other will as an object of our own willing. This is because knowledge, to the extent that it provides us with any real sense of the nature of its object, consists of ideas which are imitative representations of that of which we are thinking. It follows that if we possess a knowledge of any fullness of the willing of others, an imitation of their willing is created in our own minds. As such it must become an aspect of our own personal will, which we will want to satisfy so far as we can. True, we cannot vividly imagine all the willing in the world, but we can know that if we did, we would find in it something we would, *ceteris paribus*, wish to satisfy. From this it follows that we will see as the over-riding goal of rational existence the maximum satisfaction possible for all the wills there are. This is the basis of the construction of what Royce calls the moral ideal.

In his later ethical writings little attention is paid to this theme. He develops rather a point of view which emphasises the unsatisfactory nature of ways of life which do not give us a sense of belonging to some larger whole, more significant than ourselves. Those who realize this will come to see loyalty to some larger social unit to which we belong as the supreme value, the only restriction on proper objects of loyalty being that we should not be loyal to social units which frustrate the loyalties of others. On this basis Royce works out a morality of loyalty to loyalty. Interesting as this is, it represents a less well balanced and sensitive development of this theme, than does Bradley's more qualified version of the same somewhat Hegelian idea. It is the doctrine of *The Religious Aspect* which represents Royce's most interesting contribution to ethics.

George Santayana (1863–1952) was first a student and then a colleague of Royce's. He was strongly opposed to the absolute idealist metaphysics espoused by Royce, and had little sympathy with the doctrine of loyalty to loyalty. As a whole, his ethical point of view is very different from Royce's, but he takes a rather similar view to the early Royce on how adequate knowledge of the aspirations of others is bound to move us.

Thus Santayana holds that in grasping the aspirations of others, one will come to see their satisfaction as being as truly a good as the satisfaction of one's own aspirations. However, he differs from Royce in that while he thinks the person of insight will acknowledge that value systems other than his own have their own point, and learn some tolerance thereby, he will also be aware that if he is to live any sort of rounded life himself he must concentrate on developing the value system he has either inherited or adopted as best satisfying the vision of the good projected on the world by his own individual psyche. (In spite of his strong emphasis on the desirability of working out one's own sense of values Santayana does not favour a solitary or eccentric ethic, for by and large systems of values can hardly be lived out with much success unless they become the shared value system of a society.)

Santayana holds that what we call good and bad is that at or away from which our personal wills are directed. However, the way we actually experience our own willing is by envisaging things as possessing a quality of goodness, or in the negative case, of badness, much like Moore had in mind. This quality is cast upon the world by our own psyches, and is the aspect things wear for us, not something which holds independently; still, as such the form or quality of the good (and of the bad) are as much part of reality as are colours, which only exist for some particular type of organism. Thus the form of the good is in an important sense really there whenever anyone experiences it as there. Insofar as we understand the aspirations of others we will recognize that the good is phenomenologically present for them in places in which we do not ourselves initially find it, and if we correctly imagine the world as it is for them, we will come to recognize their way of life as having its own legitimacy. Even if in extreme cases we fight against their values for the sake of our own, we will have a kind of respect for that alien way of experiencing the world.

Reason for Santayana is present in ethics whenever we try to do justice to all the different goods which attract us and work out a stable system of values which does justice to as many different sorts of good as is possible. This effort to synthesise our varying emotional attitudes into a single harmonious, though richly variegated, approach to life is as properly called reason as is the effort to bring our varied cognitive interpretations of the world into a unity which will stand the test of experience; the one is a search for a harmony of beliefs, the other a search for a harmony of emotions.

Santayana is an extraordinarily insightful moral thinker to whom these remarks do scant justice. However, I can excuse myself in this case by the fact that I have written at some length on his ethics elsewhere. (See my book *Santayana: An Examination of his Philosophy* London, 1974, chapter X.) He is one of the main influences under which I have come to the views to be developed in Part Two.

PART TWO

CHAPTER V

Pleasure and Pain

1. The question: what is pleasure? as also: what is pain? is an important one for almost any ethical theory. Utilitarians see the effects on pleasure and pain as the ultimate determinants of right and wrong. Theorists of a more rationalistic sort are often concerned to persuade us of the desirability of promoting and preventing other things, in the recognition that pleasure (and the avoidance of pain) is the most obvious lure to human action. Moreover, any sensible ethical theorist will wish to relate his ideas to the facts of human motivation, and therefore to the correctness or otherwise of the widespread notion that pursuit of pleasure and avoidance of pain are the most basic of our springs of action. In the ethical viewpoint I shall be defending in this second part the notions of pleasure and pain will, as in the utilitarian tradition, play a basic role. Thus it will be convenient to start with some discussion of pleasure and pain.

One view of pleasure which seems sometimes to have been held, and more often to have been discussed simply in order to be condemned, is that pleasure is a certain repeatable sensation which is always qualitatively the same, although it varies in intensity.

The idea is that what is typically called a pleasant experience is a complex experience which contains the sensation of pleasure as an element (a sensation which, in principle, might be obtained on its own even if this is impossible in practice). Thus someone who finds pleasure in a certain kind of experience, whether it be listening to certain music, or playing a game, has a complex experience of the sensation of pleasure plus the experience of hearing the music or of playing the game. So when we do something because the experience of doing it is expected to be pleasant what we are really after is simply obtaining the sensation of pleasure, and we seek the rest of the experience simply as a means to that sensation, perhaps as the one uniquely effective means in our present circumstances and physical and psychological state. Whether I set out to enjoy myself at the opera, or in eating and drinking at a restaurant, I am just looking to these as presently the best means of getting the sensation of pleasure.

This view of pleasure is likely to be combined with a similar view of pain, for which whenever I seek to avoid or terminate a painful experience, whether it be toothache or boredom, it is one and the same sensation of pain which I am trying to avoid.

When this view is taken of pleasure and pain by someone who holds a doctrine of psychological hedonism, human action is seen as directed at just two ends, the obtaining of as much of the sensation of pleasure as possible and as little of the sensation of pain. Someone who goes to *Tannhauser* expecting pleasure would have gone for the sheer sensation of pleasure without any hearing of music at all, if he could have got the same amount of pleasure that way. It is basically a contingent fact that pleasure and pain are not readily obtained pure.

Even if psychological hedonism is false, as doubtless it is, it is mere common sense to allow that much of our behaviour has pleasure, or avoidance of pain, as its ultimate goal. All such behaviour, on the present view of pleasure and pain, is supposed to be directed at obtaining as much as possible of one uniform sensation and of as little as possible of another.

In principle such a view might be held about pleasure and not pain, or vice versa. In fact, it is more plausible as a theory about pain alone. However, it will be convenient to deal only with theories which take a uniform view of pleasure and of pain, leaving the reader to understand a suitable mix of comments as applying to mixed theories.

It must be understood that our concern is with theories which are supposed to explain the nature of pleasure and pain understood in the broadest possible sense, so that the words apply to any experiences which can reasonably be called pleasant or unpleasant. To call all pleasant experiences pleasures is not problematic, but to call all unpleasant experiences pains is. Even if hearing a particular boring lecture was unpleasant one would hardly normally call it painful or a 'pain', except in order to be peculiarly rude about it. It may be thought rather that the proper use of 'pain' is to refer to a particular sort of physical sensation which may or may not be unpleasant, and that many unpleasant experiences occur besides pain. However, I shall use 'pain' to refer to any unpleasant experience and only to such. This conforms to the tradition of utilitarian writing with which I shall be allying myself. However, to avoid misunderstanding, I shall also sometimes speak of unpleasant experiences or 'unpleasure'. Of course, on the present theory all these do contain a specific element of 'pain', but presumably that will not be the physical sensation of pain.

2. The view that pleasure and pain are individual repeatable sensations has been widely criticised, especially in the case of pleasure. It is also apt to strike people as an unattractive view. Sometimes it is propounded, as part of a hedonistic psychology, in an enthusiastic spirit of debunking human pretensions. It is sometimes held that the brain is programmed so that behaviour which provides the uniform sensation of pleasure is positively reinforced, and that which provides the uniform sensation of pain is negatively reinforced. (For some discussion of the views of experimental psychologists see appendix to chapter six.) Thus, we learn, the spiritual aspirant, or the 'culture vulture', with all their pretensions, are after just the same thing as the common pleasure seeker, or even heroin addict, and simply pursue it by a safer, though perhaps less immediately effective, means.

The view does seem unattractive and it is interesting to ask why. The idea that people will be leading lives as good, judged in terms of what *we, here and now*, really want for ourselves, if in the future they spend most of their time with their brains plugged into devices which stimulate their pleasure centres, and never feel the need to do anything else beyond what is needed for survival and sustaining the relevant technology, seems to make a mockery of the variety of things which we like to think of as worth doing and experiencing. But if we really do just want that uniform sensation of pleasure (and have no other desire except the avoidance of the uniform sensation of pain) what is it within us which rebels at such a picture of ourselves? The answer may well be that the main cause of our rebellion is simply our recognition of its falsehood as an account of what we want. It could be, indeed, that such an account is correct so far as hedonically motivated behaviour goes, but that we have other non-hedonic motives. But it is surely wrong even as an interpretation of hedonic motivation, whether there is non-hedonic motivation or not.

The main way of testing whether this is a correct account of what it is to be a pleasure or pain must be introspection. One must examine the various experiences which it seems apt to call pleasures and see whether one can detect some one single kind of sensation there. It seems clear to me that one cannot distinguish some separable element within different pleasant experiences which is just one and the same. It is even more absurd to think that it is some such separable element which one is really after when one seeks some specific pleasure, like that of listening to a particular symphony or going for a walk on the hills.

True, many philosophers today think any such introspective method is a quite inappropriate way to examine a question like this. How can I know that what I discover to be true about the experiences I call pleasures when I have them, is true of the experiences of other people which get described as pleasures? 'Pleasure' and 'pain' are words in the public language and their meaning could not be given by a correlation which each person can only set up between words and their own experiences, and has no basis for applying to the experiences of others. Moreover, it is thought that there is no real sense to the supposition that I could give a word a meaning by some kind of private act of correlation of this sort, for there would be no way in which I could check that I was using the word consistently, so that the very idea of so using it is without real meaning. One way or another what makes an experience a pleasure or a pain must be something publicly testable.

However, even if there were no guarantee that pleasurable and unpleasurable experience is the same sort of thing for you as for me, it would remain a reasonable belief that they are. Whether we can prove it or not, the whole of our lives rests upon the assumption that other people's consciousness is fairly much the same sort of thing as ours, – or at least mine does, and I am pretty sure yours does too. So why not deal with the problems of life, language and ethics on the basis of this assumption which informs one's whole way of 'being in the world'? So in my enquiries into the nature of pleasure and pain I shall take it for granted that there is a reasonable amount of congruence between the experiences which are associated in each of us with the same sort of public manifestations. My concern, in any case, is not with how the words have been given meaning, but with the nature, and role in our lives, of the realities which they now denote for us, when taken in the broad usage explained. The obvious way to carry on this enquiry is by an examination of one's own experiences, though one can make this more acute by what one understands others to be saying about theirs.

To hold that pleasure and pain are each of them simply a certain uniform sensation, present identically the same in all pleasures and pains, comes close to saying that there is really just one sensation one likes and one which one dislikes. But that seems quite untrue, even when dealing with one broad type of experience. For example, I like both the music of Handel and Wagner. Even if there is (as seems doubtful) some kind of common *frisson* which they both produce, that is not what essentially I like about each.

3. There is another view of pleasure which quite a number of philosophers have favoured in the last several decades. A classic statement of it is given in the chapter called 'Pleasure' in Gilbert Ryle's *Dilemmas*. [See also GOSLING.] These philosophers do not usually give the same sort of account of 'pain'. That is largely because they take 'pain' in the narrow sense; they might well say something rather similar about 'unpleasure'.

Ryle sets out to show us that pleasure is not a sensation. He gives various reasons, such as that, typically, a pleasure is not experienced in some particular part of the body. He seems to give rather inadequate attention to the many pleasures which are experienced in some part of the body, pleasures of the palate for example. However, he is right that the pleasure of going a pleasant vigorous walk, or of reading an exciting novel, are not to any great extent a matter of titillation felt in some bodily part. Another reason he gives for saying that pleasure is not a sensation (while pain is) is that one can attend to a pain and then ask for its cause (and often not know the answer). In the case of pleasure, he suggests, it is very odd to attend to a pleasure and ask what is causing it. He also objects that the sensation conception of pleasure represents it as something which runs alongside the activity, as a kind of series of thrills having nothing much to do with the activity one is enjoying.

To some extent Ryle is rightly attacking the view which we have just criticised ourselves, but in his positive remarks about what pleasures actually are he seems to throw the baby out with the bath water.

He tells us that a pleasure is an activity or occupation which we enjoy, not something other than the activity. Pleasure cannot be a distinct experience which accompanies or is caused by the activity. If it were, the more one liked an activity, the less easy it would be to engage in it, for an associated experience of pleasure would continually draw one's attention away from it. The person who enjoyed a game of chess would be distracted from the game in a way that someone who liked it less would not, which is absurd. For Ryle enjoying an activity just means having a strong propensity to engage in it, and when engaged in it to continue with it and remain absorbed in it.

This view of Ryle's, that the pleasurableness of an activity is just publicly recognizable persistence in it, seems to lose all sight of the sparkle which life has at its best. If the activity and the persistence are thought of in a behaviouristic fashion, then enjoyment can be ascribed without any real ascription of consciousness at all. Even if it is assumed that the person experiences subjectively the activity he is engaged in, it seems to be denied that there is anything pleasurable about an activity other than the amount of time he tends to put into it. This gives a strikingly joyless picture of pleasure or happiness.

4. If pleasure can be explained neither as a uniform sensation, nor as whatever we tend to persist in, let us consider a third view with some plausibility. [REM B. EDWARDS] According to this, what is common to all experiences which we call pleasures is that the person who has them *likes* them. There may be nothing at all in common with all the experiences one likes except the fact that one has this attitude of liking directed at them.

This theory may be understood in two ways. On the first interpretation, pleasures do not have to be experiences (in the sense of states of consciousness) at all. They are simply situations which are liked by the person for whom they are pleasures. A shared walk in the botanical gardens may be a pleasure for one person, a bore for another.

This idea that some merely objective state of affairs can be properly called a pleasure seems to betray the whole significance of the word. Can I have pleasures when I am dead, and assuming no after life? Certainly, I may like the idea of something happening after my death, and it may actually occur then. Is it therefore a pleasure of mine when it occurs? Surely not. True, Aristotle thought one could think of things occurring after a man's death as affecting his happiness, but 'happiness' in his sense is not the same as pleasure in ours or his, and is more tenuously related to the notion of pleasure than is 'happiness' in modern English. Besides it is not clear that he was not ascribing some kind of sentience to the dead.

It may be said that regarding a pleasure as an objective state of affairs, rather than as an experience, does not imply that anything of which I like the idea is a pleasure for me. I must be there to like it for it to be a pleasure. But it's not clear why I cannot be described as liking situations expected to occur after my death, if liking is not used in a sense which confines what is liked to something personally experienced. A reply might be: indeed one can only like what one subjectively experiences but that does not mean that it is the experience, rather than what it is an experience of, that is liked.

If the idea is to get away from the idea of pleasure as some kind of inward sensation distinct from what it is like to live in the public world, then it is well taken. If I like a walk in the botanical gardens it is that walk, as I experience it, which I like. But then the public situation as I experience it is an experience, in the sense that it is part of what someone would have to know in order to know what it was like being me at the time. One likes things as one experiences them, not as they exist in some kind of impersonally objective manner. So this view is hardly satisfactory taken thus.

But what if it is taken in a second way, according to which a pleasure is indeed an experience, but not any one particular kind of experience, or one containing any particular component, but simply any experience which is liked (that is liked at the time of having it, for the fact that I *anticipated*, or even *recalled*, it in a positive manner would hardly make it a pleasure if I did not like it at the time).

What is meant by liking? On one interpretation, it is some kind of tendency of the organism to promote or sustain certain states or processes in itself. Thus interpreted, liking is not a subjective feeling, just an occurrence in the impersonal world of supposed objective fact. Alternatively, some kind of feeling or mental act going on within the individual consciousness may be meant. (Doubtless some combination of these answers is also possible.)

It hardly goes to the heart of the matter to say that pleasure is experience which is liked, if liking is interpreted in the first way. Doubtless it is true that organisms 'go for' the experiences they like, and find pleasant, in a publicly observable way. But a creature which had experiences with a kind of unexciting deadpan quality to them, and was programmed to act so as to obtain them, so far as possible, would hardly experience pleasure.

If we are to do justice to the actual flavour of pleasure, as something we live through, we must take liking in the second way and conceive of liking itself as some kind of subjective activity. That could indeed include the physical 'going for' the thing as experienced from the inside, but then that is something subjective in the end, not a publicly observable physical occurrence. One way or another the kinds of liking in question must be some kind of subjectively lived through way of being in favour of the experience at the time it occurs.

To see pleasure as an experience which is 'liked' in some such way is better than either the uniform sensation view, or any more behaviouristic view, but it is not really satisfactory. For it implies that I can only experience pleasure when I direct some kind of separate mental act at the experiences I like. But here we may echo Ryle, though to a rather different effect. The person for whom a vigorous run is a pleasure, or a game of chess, may be so entirely involved in his experience that there is no separate act of liking or enjoying directed at it. It is, indeed, one of the paradoxes of human psychology, often noted, that one often wishes to savour an experience as pleasant and in doing so destroys its real pleasantness, while yet if one does not savour it one feels afterwards that in a sense it passed one by. There is a real dilemma here bearing on how far life should be constantly examined, one from which animals are presumably free. But to suppose that there is only pleasure where there is some kind of separate act of savouring, or some other form of liking, does not do justice to the central place in life of straight unexamined happiness (and unhappiness). Indeed at some point there is likely to be a savouring, on the part of the person who savours his experiences, which itself is pleasant or unpleasant without being the object of any further act.

So it seems that we cannot identify pleasure with liked experience, whether the liking is a physical pattern of behaviour or a subjective mental act, and it would not get over the objections merely to combine the two. We can only say that pleasure is experience which is liked, if by 'liked' we simply mean 'pleasant', which does not get us very far. Moreover, saying that pleasure is 'liked experience' obscures the fact that the pleasantness of an experience is something inherent in it, not a matter of some mental activity being directed at it which is separable from the experience itself. Where there is such a separation it is rather that the idea of the experience is pleasurable than the experience itself.

5. We need an account for which its pleasurableness or painfulness is an inherent feature of an experience, but which avoids regarding pleasure and pain as mere uniform sensations. One proposal worth considering is that pleasure and pain (in the broad sense) are hedonic tones pertaining to pleasant and unpleasant experiences, rather than separable sensations which accompany them.

For this view, experiences of all kinds may have a certain positive or negative hedonic tone. There is no special sensation called pleasure which could occur and be a case of pleasure and of nothing else. A pleasure is always an experience of some other definite sort which possesses a pleasurable tone. If I find it pleasant to hear certain sounds that is not because hearing them causes pleasure as some further distinct sensation, but that my hearing them has itself a positive hedonic tone.

The question arises whether hedonic tone is of just two sorts, the pleasurable and the unpleasurable, varying only in degree, or whether there are all sorts of qualitatively different hedonic tones. If the first, we have not moved far from the 'uniform sensation' view of pleasure and pain and there will be much the same grounds for rejecting it. Although pleasure is no longer conceived as a complete sensation which could occur independently, it will still be true that whenever we do something because it is pleasurable, it will be for the sake of feeling one and the same uniform quality. This has to qualify some experience with other characteristics than its pleasurableness, but so far as our motivation is hedonic, it will be indifferent what these other characteristics are.

Thus it seems we must suppose that hedonic tones are of many different specific kinds. Thus understood, the hedonic tone view of pleasure and pain is the most promising yet considered but it poses certain problems. Can the same hedonic tone belong to two different sorts of experience, or is each hedonic tone tied down to a particular sort of experience? If the latter, then can that sort of experience occur without that hedonic tone? Thus we might ask whether the precise hedonic tone of some particular act of hearing a piece of Handel could be duplicated in the looking at a painting by Canaletto. If not, what of two acts of, say, hearing the same musical performance? Could these differ only in that one was pleasurable and the other not, or would this have to arise from some other difference, such as a more effective registration of the musical detail?

6. We have considered four main views of pleasure and pain: (1) that pleasure and pain are uniform sensations; (2) that pleasure is whatever we tend to persist in, pain whatever we tend to avoid; (3) that pleasure is liked, pain disliked, experience, where 'liking' may either (a) be a behavioural tendency or (b) a subjective mental act; (4) that pleasures and pains are experiences with a wide range of hedonic tones, falling under the main general classifications of the pleasurable and the unpleasurable. Having seen something of the difficulties of each of these I shall now develop a positive conception on my own account which seems more satisfactory. It is nearest to the 'hedonic tone' view, is perhaps even a version of it. However, it seeks to avoid the suggestion carried by that terminology, that pleasure and pain are mere frills attached to the main body of our experiences.

A main inspiration for my view comes from Mill's qualitative utilitarianism. (An earlier utilitarian who insisted on the radically different natures of different pleasures was Hume. See *Treatise* p. 472.) That seems best understood as implying that pleasantness is a generic quality of which different pleasures are specific forms. Pleasure is related to the specific sorts of pleasure somewhat as colour is to the various shades of colour (or red to the various shades of red). Just as colour can only be exemplified in some specific version, so with pleasure. On the other hand, it is no more true that there is nothing one and the same present in all cases of pleasure than that there is nothing one and the same present in all cases of red. There is a real element of identity, but what is identically the same cannot be picked out as a separable element.

I am taking for granted a realistic view of 'universals' which may not appeal to everyone. Some may be inclined to say that there is no one and the same thing present in all shades of colour, or even in all shades of redness. They are each utterly different, having no element in common, but they fall under the same general word in virtue of the fact that a series can be produced leading from any one to the other by way of very similar qualities.

I do not think that sort of view satisfactory as a theory of universals, (or rather of generic universals and their relation to the specific universals which fall under them). But my main point is simply that pleasure is related to the various sorts of pleasure as red is related to the various shades of redness, – the correct analysis of the one case, whatever that may be, will apply also to the other. The same goes for pain, in our broad sense.

This view fits my own introspective observations. I would support it more discursively on the ground that it seems the best way of avoiding objections which are fatal to the other theories. The 'uniform sensation' view conceived all pleasure guided activity as a quest for the very same thing by different means. Whether someone goes to a concert, a brothel, or a romantic encounter, he is thought of as simply seeking a certain unvarying sensation. This sensation normally only comes to us accompanied by, and caused by, some usually complex human experience, and what the relevant associated experiences are depends on one's present physical and psychological state. But there is nothing that one is looking for, in pleasure guided activity, other than that one sensation, quite unvarying, except in intensity. This view is especially unconvincing and unattractive if combined with a solely hedonistic theory of motivation, but it remains so even if our behaviour is conceived as only partially pleasure guided.

Theorists who deny that pleasure is a sensation usually leap to the opposite extreme and say that there is no identically one and the same item called 'pleasure' which men pursue at all. To think this unsatisfactory is not to assume uncritically that a general term always picks out a genuinely identical something, but to hold that the polarity of pleasure and pain, or the pleasant and the unpleasant, would not have assumed such a basic role in human thinking if there were not one and the same recurring contrast here between the same ever recurring factors. Even those who deny that pleasure and pain are identically repeatable qualities try to find some universal account of the contrast, but locate it much less plausibly in effects on behaviour which are extrinsic to their felt character as experiences.

7. These theorists are right, however, that an adequate account of pleasure and pain must explain the central part they play in motivating action. Most theories of pleasure and pain which meet with much approval from philosophers today are very unsatisfactory in this respect, having one or other of two contrary defects. Some link pleasure and pain in some appropriate way to action, but do so merely on the basis of a verbal definition, which loses sight of their inherent nature as qualities of feeling. Others rightly take pleasure and pain as essentially qualities of feeling which we can only know from our own experience but leave us with nothing but a purely contingent relation between these feelings and our observable behaviour as creatures with wants.

Certainly psychological hedonism, the theory that a person only aims at obtaining pleasure and avoiding pain for himself, is wrong. (It is wrong even when it abandons the view that we always do what we think will yield the greatest surplus of pleasure over pain, and allows that individual pleasures and pains may concern us without even an implicit calculation of that sort.) On the other hand, it is certain that the obtaining of pleasure and the avoiding of pain is a major motive in human conduct, and one which seems peculiarly intelligible. Let us call conduct 'hedonically guided' insofar as it is motivated thus.

An individual whose conduct was in no way hedonically guided would be very odd. Even odder would be an individual who had not even any tendency to be hedonically guided, so that his not being so was not due to something like moral effort. And what of a person who was so to speak counter-hedonically guided, that is one for whom pleasure and pain changed places in their effects on his motivation, and this not as a result of some strange moral effort, but as his most spontaneous way of being? I do not think that a masochist is an example of this. For one thing, it is only in certain contexts that he even seems to be counter-hedonically guided, and even there I suggest that the so-called pain he pursues is a pain overlaid or permeated by a pervasive pleasurable quality.

8. Could there be people who were not at all inclined to be hedonically guided, or who were even spontaneously counter-hedonically guided? One is inclined to say that there is something incoherent or inherently absurd in the idea of such people.

Theories of a behaviourist tendency (like 2 and 3a above) would explain the oddity of such things by insisting that by 'pleasure' is simply meant experiences of whatever sorts people are spontaneously disposed, *ceteris paribus*, to pursue and seek to sustain and by 'pain' is meant experiences of whatever sorts people are spontaneously disposed, *ceteris paribus*, to shun and to seek to terminate. So pleasures and pains have this kind of action guiding power simply because only such as do are called pleasures or pains.

There are, in fact, several ways in which these views could be developed. (1) One might say that 'pleasures' are the kinds of experiences humans (and animals?) are typically inclined to pursue and seek to sustain and 'pains' are experiences playing the opposite role in our lives. In that case it is necessary that one who is not hedonically guided is an oddity, but not logically impossible that there should be a few such people. (2) Or one might say that 'pleasure' means the kinds of experience which, in the case of the individual in question, he spontaneously pursues and seeks to sustain, and similarly, *mutatis mutandis*, with pain. In that case a counter-hedonically guided individual would be logically impossible, and to speak of someone as merely not hedonically guided would be to say that there were no sorts of experience he was spontaneously inclined to pursue or avoid. (3) A third account of pleasure and pain along these lines might dispense with the notion of sorts of experience and simply see a pleasure as an experience which a person would, *ceteris paribus*, try to sustain if he had it and pursue if he envisaged it correctly and a pain as one he would similarly try to avoid and terminate. (Since only sorts can easily fit into counterfactuals of this nature, I doubt if there is a real alternative here.)

Of these accounts the second is the most promising, inasmuch as it would support the generalisations it seems most natural to make about human nature. But I do not think we should accept it. It leaves us with the idea that there are no real, albeit generic, qualities of pleasure and pain at all. But surely there is something about an experience being nice or nasty which, whatever some philosophers may argue, one does, in fact, know from one's own case and assume to hold in the case of others. If your unpleasant experiences have no sort of generic affinity to mine, I am completely misunderstanding what is going on in you when I sympathise with you in your suffering. Similarly when I have various attitudes to your pleasure. If you are only having experiences which, in a behaviourist sense, you are programmed either to promote and sustain, or the opposite, then nothing is occurring which can reasonably be called happiness or unhappiness and I am simply in error if I feel for you in a sympathetic fashion. If you are having experiences which you 'go for' in some more inward fashion, but they have no quality of pleasurableness or painfulness such as I ever encounter, then I do not know what your subjectively 'going for' them is, since such inward liking or disliking of an experience is itself deeply linked with finding them pleasant or unpleasant.

So I dismiss the view that it is an analytic truth (true merely by definition) that people are normally hedonically guided in their conduct to a great extent (at least if the analyticity arises in that way).

Such is the (continuing) prevalence of the view that the only really strong kind of necessity is logical or analytic (not seriously dented by some talk current about natural or even metaphysical necessity) that those who accept that pleasure and pain are distinct (though generic) qualities, are inclined to deny that it is necessary that people, other things being equal, seek to pursue and sustain pleasure and to avoid and terminate pain. Such people will think it a mere deeply entrenched, but contingent, feature of human nature that pleasure and pain play this part in our lives.

This seems wildly implausible. Moreover, it has the difficulty, seized upon by the behaviouristically inclined, that it is most unclear in that case that I have any right to assume that it is in fact pleasure and pain as I experience them which play that kind of role in the life of others. Perhaps I am the only one who goes for pleasure as the quality I identify in my experience, and turns away from pain understood similarly.

The only way out of this dilemma is to accept that some things do of their very natures tend to have certain sorts of effect, and to be resolute in accepting two consequences of this (1) that there is a real nature involved here, to speak of which is not simply a way of talking about the effects, regular or usual, of what possesses it, but something which can be identified as a quality present in the thing itself, (2) that this tendency has a recognizable necessity to it.

To accept that this is so, is to turn away from that denial of genuinely intelligible necessities in nature stemming from David Hume, and still deeply embedded in the minds of philosophers of an empiricist persuasion, to the natural expectation of the human mind that reality is intelligible, and that it is of the nature of certain sorts of things to exert a certain sort of influence in the world. A.N. Whitehead suggested that the Humean denial of real necessity arose from a concentration of attention on the visual, where such necessities are absent. We find them, claimed Whitehead, at a level of experience which he called 'perception in the mode of causal efficacy', consisting, roughly, in bodily and emotional feeling. [WHITEHEAD pp. 172 et ff.] Though I do not accept the details of his account, there is certainly something in it. The successive arrays of objects in visuo-tactual space can be imagined different in all sorts of ways, and it will always seem intrinsically possible for us to perceive external occurrences which do, or seem to, go against the most firmly established laws of nature. It is only when we come to the emotions that we enter a realm of intelligible necessity. (In the end this is one reason why I join Whitehead in thinking that nature at large has an inner emotional life, for in the light of this we can take the laws of nature as a registering from the outside of what is necessary from within.)

9. Thus I think that we should not be afraid of the idea that pleasures and pains are of their very nature liable to affect behaviour in certain directions. And the truth of the matter seems to be somewhat as follows. The pleasurableness of an experience tends of its very nature to promote activity within the stream of consciousness which tends to sustain and repeat it, while the painfulness of an experience tends of its very nature to promote activity which will remove it. In some cases the effects of these activities on the sustaining of pleasure and removing of pain need to have been exhibited in past experience, in other cases, conceivably, this is not so. I shall develop no view as to the balance between acquired and innate linkages of this sort, beyond remarking that where the association between pleasure or pain and activity has been set up only as a result of past experience, we are to think of this as a causal process rather than one in which the mind learns truths about what sustains or removes certain pleasures and pains.

If these effects within consciousness are to influence the behaviour of the organism as a physical reality, then pleasure, pain, and the activity which they promote within consciousness must be somehow intimately related to the physical activity which tends to promote or remove the physical bases of pleasure and pain. How this is to be understood constitutes the kernel of the brain/mind problem with which we cannot deal here. My own inclination is to think that the core activity in the brain which performs this job is what the actual experiences of pleasure and pain and the mental activity which they promote present themselves as being to the neurophysiological observer. In their own being they are experiences but they register themselves in the public physical world as events in the brain.

To say that pleasure and pain have, as a matter of necessity, the tendency to influence behaviour in this way is not to say that they are objects of deliberate pursuit. However, it suggests an explanation of how they can become such, and are bound to remain peculiarly basic goals.

I suggest, first, that when the pursuit of goals is of that deliberate kind which goes with some conscious envisagement of that at which one aims it is caused by a feeling of desire. But what is a feeling of desire for something? My provisional suggestion is that it is a certain sort of pleasant idea of it, and the desire for something not to occur is a certain sort of unpleasant idea of it. (Its distinctive nature is that the states of affairs they envisage are presented as bathed in the pleasant or unpleasant quality of the idea and as something realizable or preventable by my action.) It follows that desires, as themselves pleasant or unpleasant experiences, will promote activities which sustain them in being and strengthen them. Nothing can do this so effectively as evidence that the ideas are, or are about to become, true. So positive desires tend to produce activities which bring about the situations of which they are the desire and negative desires do the opposite, and this is a consequence of the intrinsic efficacy of pleasure and pain in promoting what respectively sustains or removes them.

I am far from maintaining that all our desires are for pleasure or the avoidance of pain of some sort. Other things besides these can present themselves drenched in the pleasantness or unpleasantness of our idea of them, and as realizable or preventable by our activity. All the same, there is bound to be a certain primacy to pleasure and pain as objects of desire, since the actual character they will possess if they occur corresponds in a way in which other desired objects, on the face of it, cannot, to the idea we have of them in desire. So granted the special efficacy of pleasure and pain as causes, they are especially calculated to become objects of desire.

Another way of putting all this is to say that an individual consciousness of its very nature as-it-were strives to sustain and repeat pleasure and to terminate and prevent pain. This as-it-were striving is, in its most basic form, not a conceptual affair involving any sort of judgement or knowledge of fact. It is simply a kind of primitive struggle to sustain pleasure and terminate pain and to do the same with experiences which have been associated with pleasure and pain in the past, without any judgemental memory of this. Upon this basis grows up the tendency of pleasant ideas of situations to bring about the situations they envisage, and the opposite with unpleasant ideas. It grows up because the primitive struggle to sustain pleasure and terminate pain leads to activity which will make the pleasant ideas true, and therefore sustainable in consciousness, and the unpleasant ideas false and therefore expugnable.

10. It must be emphasised that the pleasure which consciousness necessarily as-it-were strives to actualise within itself may be 'objective' pleasure every bit as much as 'subjective' pleasure. For one's consciousness includes not only one's body and one's persona or self as they figure in one's personal experience of the world, but also the external world as it also so figures. The flowers I look at in the garden, as they are for me, are as much contents of my consciousness as is my body, and even as is my most inward selfhood, and their pleasing quality is a species of pleasure belonging out there rather than in here. It is not that I register the shape and colours of the flowers and feel an inward *frisson* in response, but that the flowers themselves are directly pleasurable. The direct pleasurableness thus found in the external world, as it figures in our experience, is as much something consciousness strives to actualise within itself, as any pleasure felt in the body or anything especially pertaining to the self.

George Santayana, in his *The Sense of Beauty*, described beauty as 'objectified' pleasure. The idea was that the experience of beauty was the experience of our own feeling of pleasure in response to certain objects, mistakenly taken as pertaining to the nature of the object. But, for reasons Santayana might have approved, it is better to call it 'objective' pleasure.

To talk of the pleasurableness which is an inherent part of the objects we see around us as objectified pleasure suggests that it was originally experienced as a sensation within the subject and then somehow projected onto the seen objects which produced that sensation. But certainly we do not at present perform any such act of projection. There is nothing on which we could even try to project the pleasure other than that which is already its locus, the objects seen around us. Nor is it reasonable to think that an act of projection occurred in infancy. It is more reasonable to think that infantile experience begins with no clear division between self and not-self, but that gradually the contents of consciousness are sorted out into what is experienced as self, mainly one's own body, and what is experienced as not-self, mainly the experienced surround of one's body. Pleasure and pain belong equally on both sides of the division.

If there is anything by way of projection it is rather a projection of the pleasure present in the surrounding world into oneself. The not-self (including an objectified version of one's own self) is supposed to be shared with others, and one's awareness of others is bound up with taking one's own experienced not-self as something which figures also in the not-self of other consciousnesses. [Cf. A.E. TAYLOR Book IV Chapter 1] The objective physical world of a slightly sophisticated common sense is a kind of highest common factor of the not-self aspect of human consciousnesses in general. Since I find that the pleasure and pain which is there in my own not-self is often not there when it figures as the not-self of others, I tend to treat them as not really belonging to the objects but to my response to them. Thus I introject the pleasure into myself. But it is not very clear into what I introject it. It is certainly not felt as a feature of my own body. (The beautiful sound of the clarinet is out there, not within my body.) And it does not make much sense to say that I introject it into my own consciousness, for that is where it is already alongside all the things I perceive in the immediate character they wear for me. The problem springs from the difficulty which common sense metaphysics has in keeping straight the distinction between my consciousness, on the one hand, and my self and body on the other.

Doubtless the assertion that beauty of certain sorts is subjective serves fairly well as a way of saying that the beauty differs much more from one person's personal version of the public world than do other qualities. The presented hedonic quality of things is less of a shared possession than are shapes or even colours or even the gestalt characters of every day objects. But our whole conception of what is *really there* in the public physical world is a makeshift affair oscillating between incompatible elements according to the needs of the moment. I am sure that there is an ultimate system of things in themselves behind the world as we experience it, but that is something quite different from the world of daily experience and may be set aside for now.

11. We decided that pleasures and pains are of radically different and incommensurable kinds, though all are specific forms of the identical generic qualities of pleasure and pain. There is

something one and the same in the pleasurableness of eating chocolate, in the peaceful charm of a running stream as we see and hear it, in the sense of fulfilment and of enhanced personal worth felt in the successful achievement of some physical or mental feat involving the overcoming of more immediate desires for comfort, and in various forms of sexual excitement, but this is not a matter of some detachable element of mere pleasure which accompanies what is distinctive in each of these experiences. That being so, we cannot raise the question whether consciousness is steered towards some maximum of pleasure, or some greatest possible surplus of pleasure over pain, for it is none too clear what these quantitative expressions would mean. But even if pleasures and pains cannot be *summed*, perhaps there is still such a thing as a greater and lesser degree of pleasurableness to consciousness as a whole (to which individual pleasures and pain contribute) and some tendency for consciousness to do what is required in order to keep its general pleasurableness as high as possible.

None of this is to say that what we are constantly after is pleasure and the avoidance of pain for ourselves. The striving of which I have spoken is a teleological tendency not an object of will. Will, as I see it, consists in the control of action by compelling ideas of the future. These compel according to their own degree of pleasantness or painfulness, but the goal of the action they produce is the actualisation or non-actualisation of their content, that is, of that of which they are the idea.

12. Our view is that there are indefinitely many specific forms of the generic universal of pleasantness, just as there are indefinitely many specific sorts of blue. Presumably just as each shade of blue is itself a repeatable universal, so is each specific sort of pleasure. This raises the question, however, whether the same specific sort of pleasure can reoccur as a feature of otherwise different experiences, or whether they can only reoccur alongside the totality of what one would typically report upon as having been a pleasant experience. Thus even if the specific sort of pleasure we can have in eating and in listening to music is different, is it also bound to be different whenever we taste a somewhat different

taste, or can two different tastes provide the same specific form of pleasure?

A closely related question is this. Should we think of the specific universals which fall under the generic universal of the pleasant as consisting in the whole character of individual pleasant experiences, or in some distinct element within that total character? Is the specific form of pleasantness a distinct feature present in the character of the whole experience as the specific shade of blue of a blue object is, or is it rather the character of the whole experience which is a species of pleasantness? Is the whole character of the experience I have at a particular moment of musical enjoyment a specific form of pleasantness or is that something contained within it?

There are certain difficulties in either answer. If we say that there is a distinguishable element of pleasure, we seem once again faced with the suggestion, even if in a less radical form than previously, that so far as life is hedonically guided what we are really after is something to which most of the concrete fullness of life is only a contingently required accompaniment. I go to a concert to get the specific pleasures which, say, Sibelius's seventh symphony gives me, but if I could get them without hearing the music that would be just as good. This view is not merely unattractive, but false. One cannot peel off the pleasurableness of certain sound combinations and think of it as something which could remain the same on its own or in other combinations.

Yet the more attractive option, that it is the whole hearing of the symphony (or each individual moment thereof, with its own particular perspective on the whole) that is the specific form of pleasurableness, is somewhat problematic. For surely someone can hear the same sounds, structured even in the same way on just the same basis of musical experience, and not experience pleasure, which suggests that the specific pleasure is a distinct and perhaps separable element. However, it is not so clear that there could be another experience the same in every respect except for the absence of that specific pleasure. Does not the precise way in which I structure the contents of an experience go hand in hand with its pleasurableness?

Perhaps the best view on this difficult matter is that there is a distinguishable element of pleasurableness within the whole experience, but that this is a pleasure of a quite specific sort so intrinsically bound up with the rest of the experience that neither could exist without the other.

13. My most controversial claim has been that pleasure and pain are distinct qualities of experience which necessarily tend to influence behaviour in certain ways. Some philosophers will object to this on the grounds that the very idea of such a necessity is incoherent. They will think that if 'pleasure' and 'pain' are really used to pick out certain qualities of feeling, then it cannot be a necessary truth that they tend to affect behaviour in certain ways. That could only be so if the expressions refer in some way to whatever types of experience tend to have such effects.

The nature of necessity is a large issue with which it is impossible to deal here. Some contemporaries speak of metaphysical or natural necessities which are quite different from analytic ones. I am not convinced that they are really acknowledging the existence of a necessity which goes clean counter to the view empiricists owe mainly to Hume, that distinct existences cannot have necessary relations one to another, but if they are, I am glad of their support. My own view is that there are such necessities, at least of tendency, and that the reinforcing powers of pleasure and pain are conspicuous examples of such. If we deny that these, in virtue simply of being the specific qualities they are, have an intrinsic tendency to influence behaviour as positive and negative reinforcers in the way we have roughly characterised, we must either analyse them behaviouristically or pretend that there would be nothing intrinsically odd to counter-hedonically guided behaviour. We will also continue to find the 'other minds' problem unresolvable because we will not be able to see behaviour as intrinsically fitted to express modes of consciousness which are, all the same, distinct from it. True, pleasure and pain must act through the brain and the nervous system and the precise way in which they have their necessary influence cannot be understood except on the basis of some adequate view of consciousness and brain. However, to be adequate, it, in turn, must allow for the element of necessity here.

That the role of pleasure and pain as positive and negative

reinforcers of behaviour has a certain necessity to it has bearings on the 'other minds' problem which I cannot develop here, but the astute reader will realize that this does go some way to allowing that appropriate behaviours are essentially fitted to be criteria of their presence, without prejudice to their identifiability as distinctive genera of feeling.

CHAPTER VI

Desire, Will and Moral Judgement

1. Accounts of moral judgement which do not explain its special power to motivate are unsatisfactory. True, the moralist who despairs at times of mankind, probably including himself (if he is not a hypocrite) may deplore the limited extent to which moral judgement motivates. All the same, there is something decidedly odd about someone who grants that an action is bad but is not remotely motivated to attempt avoiding it. That this is paradoxical is not because we ascribe a desire to avoid bad actions to men as something extra to the judgement that they are bad. Rather, the judgement that they are bad, when sincerely made, seems of itself to involve the desire to avoid them, or discourage others from them. Thus moral or other value judgements play essentially the same role in the explanation of conduct as desires do. They are not thoughts about the means to something which is independently desired. There is no real recognition that something is good or bad unless it embodies some degree of motivation to pursue or shun it, do or refrain from it.

That it recognizes this is the strength of the attitude theory of ethics; that it does not is the weakness of most forms of objectivism. On the other hand, the attitude theory of ethics scarcely does justice to the fact that moral and value qualities seem to belong in the world as actual features of it on which moral and value judgements report. Sometimes it recognizes this, but dismisses it as an error, but this is premature until the possibilities of taking this common way of experiencing the world more seriously have been adequately explored.

An objectivism which could explain how the recognition of the presence, or indeed the mere belief in the presence, of value and moral qualities was inherently motivating would be able to associate judgements about their presence with desire in a way which was not merely contingent – something which might well have been otherwise – just as does the attitude theory, but would not, as it does, dismiss the belief that these are real features of the world. It would thus have the main virtue of an attitude theory while being nearer to the common sense of mankind. Not that common sense is by any means necessarily or always right, but, other things being equal, its approval of an idea counts in its favour.

Some will think that it is with will rather than desire that moral and value judgements need to be associated. They may think that the judgement that an action is wrong is not so much associated with a desire to avoid it as with the intention of avoiding it, intention being rather a foretaste of will than a desire. They may think that the most important role of the judgement that an act is wrong is to set our will in a direction which is opposed to desire. However, as I see it, these philosophers use the word 'desire' in a restricted way in which it stands only for desires, in a larger sense, which are of an especially sensuous or selfish nature, and as such are regarded by these thinkers as less lofty features of the human endowment than the direction of the will in favour of the morally good. This surely disguises the common structure which underlies deliberate behaviour.

2. I shall now sketch a rough and ready account of the relations between, on the one hand, moral and other value judgements and, on the other, desire and will. This will somewhat develop and, at least verbally, modify the initial account of desire given in the last chapter. Although the account certainly oversimplifies the facts I believe it sufficiently near the truth to point the way to the proper grounding of ethics. It is based on considerations which combine the phenomenological, the discursive and the metaphysical. Thus it accords with my own attempt to introspect what is going on when I will, desire or make such judgements and with what I find myself bound to imagine if I try to imagine what these are like for others and it coheres well with what seems to me the best view of the relations between the mental and the physical. Finally it provides an intelligible interpretation of our sense that moral and other value judgements are both genuine judgements about the presence of certain features in the world

and mental states which are inherently motivating. Thus it is a hypothesis, to be tested by its adequacy in dealing with the various problems about ethics encountered so far in this book, by the persuasiveness of the metaphysics with which it is associated (of which however there will only be slight indication in this work) and by its accord with each person's own experience and sense of the experience of others. (There is a brief discussion of the ideas on these things of experimental psychologists in the appendix at the end of the chapter.)

Desire is always desire for the presence of the good or for the absence of the bad. That is not to say that we desire only some abstraction called the presence of the good or the absence of the bad, but only that the object of desire is something the presence of which is judged or found to be good or the absence of which is judged or found to be bad. The same is true of the wish for something. The main difference in meaning seems to be that the word 'wish' is especially appropriate when we are not in a position to bring the thing about, or when countervailing desires stop our doing so. Thus by a wish is often meant an ineffective desire.

In speaking of desire for the presence of something, I am not meaning to distinguish the something from its presence. If I desire to see my friend drink some wine and get jollier my desire is for his drunken jollity. I speak of positive desire as for the presence of something to balance the fact that negative desire is for something's absence. By presence or absence is meant presence or absence in the world, not necessarily presence to, or absence from, my own awareness. I can desire an object which lies beyond my own life span. Perhaps my use of 'desire' stretches ordinary usage, but it seems the best way of referring to a basic phenomenon.

Some philosophers have put great emphasis on the fact that we do not simply wish for or desire an object or state of affairs, we wish for or desire it under some description. I may desire that object, or the state of affairs of my eating that object, under the description of an apple, but if I knew it was poisoned I would not desire it, so that there is a true description of it under which I definitely do not desire it. If I ride a little roughshod over this, it is partly in the interests of simplicity, partly because I think that what is really desired is best thought of as the possession by reality of some property, and saying that one desires to eat that apple is saying that one desires reality to have the property of my eating an apple of a certain sort standing there just before me, a different property from that of my eating a poisoned apple.

If something is desired, and the belief is present that some sequence of events involving my own activity will bring it about, then, other things being equal, that activity is liable to follow, unless checked by contrary desires or other desires which are at the moment more pressing. This can only happen, of course, when it is the the desire of an organism with a suitable physical make up, and in a suitable physical state. (Or at least this is the normal and familiar case, to which we may confine ourselves.) Moreover, many factors contribute to determining the precise times at which desire will spill over into action, including processes in the brain (some of them constituting so-called unconscious desires, others various kinds of phasing mechanism) which have no obvious analogue in the main stream of consciousness other than this inhibitory effect upon the efficacy of the desires there. Still, it does seem that desire has an intrinsic tendency to spill over into action in this way. Such activity constitutes *deliberate behaviour*, which consists essentially in the movements of an organism under the influence of desire.

Of course, in some circumstances desire may produce rather the inhibition of overt behaviour which would otherwise have taken place under the influence of other desires or in some more instinctive manner. It may also produce shifts within the stream of consciousness (and their physical analogues in brain process) without direct issue in action. However, there is no sharp divide between the case where the effects of desire are simply on consciousness and when they are on the physical world. When desire produces shifts in consciousness it produces changes in the brain, and when it issues in behaviour it issues at the same time in the conscious registering of that behaviour.

Will is simply desire insofar as it is efficacious. I will to do that which is produced by desire in the fashion described. Intentions made in advance are desires in the early stages of their efficacy, or in the actualisation, which may never be completed, of their tendency to be efficacious.

There is much behaviour which is not fully deliberate. Much of the time we get on with life on the basis of instinctive or conditioned responses where there is little or no consciousness of an end in view. However, when we are thus on auto-pilot we would hardly be behaving at all, if our activity was not quite readily accessible to the influence of desire or will, and broadly in line with the occasionally envisaged purposes of that dominant consciousness which is us. Thus in explaining how moral judgement relates to desire we explain how it relates to us as the conscious agents to which it must primarily address itself.

Desire is for the presence of something judged or found good, or for the absence of something judged or found bad. But how is the judging related to the desiring? I answer that they are one and the same thing. When I desire or wish that something shall not happen I just see its occurrence as something bad, or, if you prefer, as something that would be bad. (When something, which may or may not happen, is conceived under the aspect of the good or the bad, we may say either that we are thinking that it is good or bad, so that its actuality would be the actuality of something which, actual or not, is good or bad, or we may prefer to say that it *would be* good or bad if actual. I shall not attempt consistency on this.) To be somewhat more precise, by a desire is meant a thought that something which my action might help realize would be good (especially in contrast with the badness of the present situation, conceived as lacking it) or that something which my action might help prevent would be bad, when this recurs in consciousness with sufficient persistence to be able, other things being equal, to affect action. Moreover it is not merely some special class of value judgements which can thus become desires, but any value judgements which ascribe genuine value, moral or otherwise, to anything.

3. Some philosophers have assimilated value judgements and desires as a way of dispensing with any such qualities as good or bad. It should be clear that their assimilation here has no such aim, since I am claiming that a desire is a judgement about the presence of these qualities in some state of affairs, and will be going on to say that such judgement can be true or false.

We have prepared the ground for our view of what these qualities are in the last chapter. Goodness is something we actually encounter in certain states of affairs as we directly experience them. It is the attractiveness, or sparkling quality, which permeates what we like experiencing in the immediate experience of it, while badness is the forbidding quality which permeates what we dislike experiencing in the immediate experience of it. As such they are the same as pleasure and pain, or unpleasure, taken in a broad sense. For there is no genuine distinction between saying that the direct contents of experience are good or bad, and saying that they are pleasant or unpleasant (provided of course we distinguish in the latter case between pain or unpleasure, qua the unpleasant in experience, and pain as a physical sensation which may or may not be so).

Thus when we think of some state of affairs, which may or may not occur, as good or bad we are thinking of it as having something generically in common with good or bad, pleasant or unpleasant, experiences or objects of experience. (This may not apply to what we merely think of as good or bad as a means, but then there will be something to which it is supposed to lead which will be thus conceived. Not that this distinction should be pressed too sharply for the felt quality of the end soon permeates the means to it as we conceive or even as we experience it.) This holds, in a way I shall explain shortly, even when we would positively deny that we are thinking about it as an element of anyone's experience.

4. For I am not in the least denying that we think of many things as good or bad (in themselves, and not merely as a means) without thinking of them as wholly or at all consisting in pleasant or unpleasant experience. There is, indeed, no end to the variety of things which people may think good or bad. I may, for example, think it bad that my university library is buying so few philosophy books, that children are watching video nasties, that the great rain forests of the world are being steadily cut down, that people are being tortured in the prisons of various countries, that the government is insensitive to the unemployed, that so many people are unemployed, that Smith is gloating over the fact that he has defrauded Jones, and so on. My suggestion is that these situations are envisaged as having those same attractive and repellent qualities of which the paradigm examples are the pleasurableness and unpleasurableness of what is directly given in experience.

At a rather different level of desire and valuation, I may think it bad that I should not be paid as much as someone else, that people do not seem to respect my judgement, that I should not be invited to one party, and that politeness requires me to attend another one. I may not apply such words as 'bad' to these situations in public discussion, since, as we shall see, this would suggest that it could be taken up into one of those shared valuations which pertain to our construction of a common world and system of values. But I may still envisage them as possessing that quality which is, in fact, what I have in mind when I think of any situation whatever as bad.

The thought that these situations are good or bad will influence behaviour as a result of the same basic mechanism as that in virtue of which judgements about my pleasure and pain do. They themselves are experiences possessing something of that positive or negative value charge which they attribute to what they envisage and as such will promote activity which enhances or negates them by forwarding or hindering their objects. How we manage to think that these qualities can exist as features of such situations and thus can occur other than as the pleasurable or unpleasurable quality of experiences will be considered shortly.

If we recall the lessons of the last chapter we can see in rough and ready terms why desires, understood as judgements about what is good or bad in this basic sense, influence our behaviour as they do. At that point we were concerned only with the influence on conduct of anticipations of pleasure and pain for myself or others. But now that we see that the goodness and badness, which we envisage as pertaining to that whose presence or absence from the world we desire, is in fact always a species of those qualities which, occurring in experience, constitute pleasure or pain, we can see how the same basic mechanism gives all desire its potency. To the extent that it is a genuine conscious fact it will itself possess something of the quality it attributes, the more so the more fully I really bring to consciousness the nature of what I am after, and thus be a pleasant or unpleasant experience. As such, it must like any other pleasant or unpleasant experiences promote activities which sustain and strengthen it if pleasant or which weaken it if unpleasant. Nothing can do this so effectively as evidence that the ideas are, or are about to become, ideas of what is or is not actually present in reality. Thus the tendency of desire to lead to action which realizes its object is a consequence of the intrinsic efficacy of pleasure and pain in promoting what respectively sustains or removes them.

Of course, the thought which guides action is often almost entirely verbal and, as such, it may seem to involve no real envisagement of the goodness or badness of what we are seeking or avoiding. However, words themselves tend to occur in consciousness with a hedonic quality which reflects the value ascribed to what they specify as goal, means, or upshot to be avoided. Besides, unless one's consciousness is very impoverished, purely verbal planning is a mechanism in the service of periodic desires in which one's goals are envisaged in some fuller fashion. This mechanism can have a kind of fine tuning to the details of the action required which more imaginative thought may lack (though the converse can also be true). We should remember also that as the situations we seek to realize or prevent are themselves likely to be to a great extent verbal in their very nature (things being said, heard, read or written) their full envisagement must itself often be verbal in nature.

5. If all this is so, then the strong antithesis between desires, wishes and willing, on the one hand, and beliefs or judgements on the other, on which the attitude theory of ethics rests, is a mistake. The former are all types of judgement. They are judgements that certain perfectly genuine qualities exist as features of certain situations. As such they can presumably be true or false. Moral judgements and other value judgements are indeed desires (or at least the kinds of things which can become so) but they are, nonetheless, judgements which can be true or false.

There should not be much too much difficulty in granting this, so far as it is directed at our own future experience. If I desire the experience of seeing and hearing a certain operatic performance this will typically take the form of envisaging it as a good, that is as a pleasant, experience. If when I attend the opera, the experience is not a pleasant one, the envisagement of it which was my desire to have that experience was incorrect. It misrepresented what that experience would be like.

Action caused by desires for certain experiences on our own part are typically actions caused by envisaging these as having certain hedonic qualities which, in the event, they may or may not have. Surely the same is true for the most part when I desire that someone else should have, or go on having, a certain experience or that they should not have, or cease having, a certain experience. In these cases I typically envisage their experience as good or bad, as pleasurable or unpleasurable. In doing so I will be judging things truly or falsely.

Thus I may wish that someone would read a certain novel, envisaging their reading it as the same kind of pleasurable experience which I had. But in fact the experience they will get may be rather unpleasant, in which case what I think to be good will not be good. Or, to put the same thing differently, the experience which will occur if they read the novel will not be that good sort of experience which I thought would occur, but a different, and bad experience. Whichever way we put it, our thoughts are straightforwardly true or false.

But there are many desires which it would seem very odd to equate with true or false judgements about the presence in certain situations of qualities of the same general sort as the pleasurableness and painfulness of experiences. This is true not only of judgements about the goodness or badness of situations thought of as existing objectively, apart from any particular person's experience, but also of cases where we think it good that someone should have a nasty experience or bad that he should have a nice one.

6. Moreover, however the connection between desires and value judgements is conceived, it may seem quite wrong headed to think of basic value judgements about what is good and bad along the lines proposed.

It is, after all, quite common to think it a good thing that someone should suffer for something wicked they have done. In such cases what presents itself to me as a good thing is their very suffering, their having an experience, which, from their point of view, is thoroughly bad. I would not think it nearly so good if their experience had the kind of felt goodness which we have been claiming that a positive valuation of something really concerns. We should bear in mind too that there are self reproachful people who think it would be a good thing that they suffer more than they do, and not necessarily in a way involving some kind of masochistic enjoyment.

Nor is it uncommon for people to think of 'objective' states of affairs as good or bad in ways which cannot be reduced to a judgement about the experiences they produce. Take the following trite example. I think of it as bad that the university library is not buying certain books. Now my thinking this situation bad may link up in some eventual way with my thinking of it as having deleterious effects on the happiness of various (roughly specified) people. Still what I am primarily thinking bad can hardly be said to consist in anyone's having an experience. I may even think that the situation would have been bad, even if neither I nor anyone else had ever become aware of it, perhaps as even worse since then the empty shelves would have been compounded by empty heads.

The situation I envisage as bad in such a case is in a broad sense a physical one. However, it is not a situation which could be described in physics or chemistry. It is a cultural situation, for books and libraries are cultural entities. As such, it is, I presume, uncontroversial that it could not exist in a world without consciousness. All the same, the holding of such cultural situations cannot be identified with any phase or immediately given object of anyone's consciousness. Thus if I think of a cultural situation like this as bad it must be that I suppose it to possess a badness quite different from any unpleasant quality of some experience or content of experience.

Thus our whole view both of desire, and of the judgements of value with which we are equating them, seems to clash with evident facts about the range of objects of desire and the kinds of things we can think good and bad. To resolve these difficulties in our account we must go a little into metaphysics and epistemology.

7. Each of us makes a distinction between features that some part of the world wears simply for himself, and features which it possesses in some more objective way. For example, I recognize that the blurry quality of a page of print as it appears to me when I do not have my reading glasses on, is a feature which that page wears for me, and which it will not do for another, or for myself when I find my glasses. Moreover, I think that the clearly

articulate letters and words which it presents to one in a condition to read it easily constitute a feature which is really there in a sense in which the blurry aspect it wears for me is not. If I am similarly defective in other respects (as compared with the norm or with some ideal of appropriate perceptual power) such as in the colours I see, the sounds I hear, and so forth, I think of the world as wearing an aspect for me which fails to live up to the features which are really present there in the objective world. And quite apart from defects, I distinguish between what is really there in some steady fashion in some bit of the objective world, conceived as not changing in any such way as this, and the changing aspects things wear for me according to the conditions in which, or the position from which, I observe it.

I am not suggesting that we have some consistent plan, always in use, for making the distinction between features which are present in the world only for us, and features which are there in some so-called objective way. Do we think that as lighting conditions change, the colours remain the same, but the look they wear for us changes, or do we think that things are changing their colours? If there were no such thing as night time, I would think paradox was most reduced by thinking that things changed their colours as the light waned or changed. But following that line of thought, one would have to think that things had no colour in the dark, which clashes with the best interpretation of how one is thinking when one gropes one's way about in the dark on the basis of a conception of what is there which is mainly visual in character. So it seems that somehow we think of the world as consisting of shaped and coloured objects, where the colours are the very colours which would feature in our experience if we saw the things in broad daylight. (Not that this ever specifies an exact shade, or that we could remember what it was, if it did.) However, to think of it thus is to think of it as existing with features which it could not possibly have as something existing apart from consciousness of it. We cannot really suppose that things having the fairly homogeneous coloured surfaces we see them as having are genuinely there in some absolute way. Even a little thought about the more basic physical articulation of matter should show this. My remarks will be challenged by some philosophers, but if they address themselves to the heart of the matter, rather than fuss over slight nuances of expression, it should be granted by most sensible people.

In general, then, we cannot produce any consistent criterion for distinguishing the features of the world which we think of as being 'subjective', in the sense that the world only wears them for us, and the features which are 'objectively there'. Even if we could formulate such a criterion, the things which it classified as 'objectively there' would not be there in any way genuinely independent of the human way of perceiving, experiencing, and conceptualising the world. Since nonetheless we make, and have plenty of reason to go on making, the distinction (drawing the contrast in slightly different ways as suits present convenience) the best way of understanding it is to think of the so-called objective world as a kind of common construction we jointly make and in terms of which we commonly conceive our interactions with each other and our shared environment (whatever that may be in its own ultimate being) especially those which generate those very subjective perspectives each of us has on the world which are our own personal way into the shared construction.

The objective world which we construct is not, as I see it, something standing in sharp contrast with the immediately given contents of experience. If it were it would be much more of a mystery for us than we commonly find it. The case is rather that it is composed of a selection, somewhat idealised, and altered sometimes in the loci to which they are ascribed, of those contents. That is, what we think of as existing when we think of the objective world as existing is a system of things of much the same sort as those of which the sensory contents which we actually experience are supposed to be (perhaps subjectively modified) components or features.

This is not to deny that the contents owe much of their character as they figure in mature experience to the fact that they are conceived as parts of such a system. Still, the system was originally constructed by each of us, in conditions determined in part by the culture in which we grow up, out of primitive versions of those same contents.

Thus each of us contrasts the world as it is for him with the world as it is supposed to exist 'objectively', yet could not think of the objective world at all unless elements of what is directly present to his consciousness were thought of as part of it, and it as an extension of them. The world as immediately present to me is conceived of as a personal version of the real common world and I use language to express, and share in, and to some extent control my actions by, my conception of what that real world is. However, the contrast between the objective shared world, and the world as it is for me, is a subtle and ever shifting one, since I cannot engage in the constitution of the shared world without that world becoming also my personal world.

Reflective people, of course, may come to contrast the world as it is not just for them personally, but for humans in general, with the world as it is in itself. At least when this contrast is drawn in a metaphysically unsophisticated way, the real objective world is still seen as composed of features which coincide partly with what is immediately present to or in the perceptual consciousness of humans. As such, it is doubtful whether it is anything more than a further development of the cooperative construction by which we produce a shared world for human beings. Thus, in most ordinary contexts, the world as it 'really' is, as opposed to the world as it is for me, can be identified with the world as it is for humans in general.

8. For those who accept a view along these lines, judgements about what is there in the public constructed world are not true or false in the same literal way as judgements about what individuals personally experience. If I have some fairly specific thought about the character of your experience, then your experience is either something there in reality with the character I think of it as having, or I have got matters wrong in some very basic way. If, on the other hand, I judge that some feature is there in the so-called objective world, my judgement is true or false only in a more pragmatic way, perhaps as cohering or not with what others will or would construct on the basis of their experience, perhaps as what should be a helpful move in the building up of the construction, whether others are sensible enough to go along with it or not, or perhaps in some other way.

So judgements about the experiences of others, or indeed about my own past or future experiences, are susceptible of a more deeply literal truth, in which the real character of what exists in the world is captured, than are judgements about the ordinary objective world. There are, however, metaphysical theories which seek for an account of the environment which backs human communication and grounds human perceptual and bodily experience which will not merely have validity as a useful construction, but which will possess the same kind of literal truth as do judgements about our experiences. For surely there must be such a literal truth. There must be a reality, and thus a literal truth about it, which constitutes that real environment which produces those subjective features of the world in our consciousness, one in which we, whatever *we* really are, are really existing and coping with via our joint construction. To deny that there is, is to believe that human consciousness floats in a kind of void, in a manner which cannot be taken seriously.

One view about that real world, or the world in its real character, which we are in the midst of, is that it is the world as specified by physics in those parts of it, present or to come, which will best stand up to appropriate testing. I endorse this view if the relevant physics is interpreted as only telling us something very abstract, although extremely detailed, about the structure of the world. But what has this structure must have it in some concrete form. My own view is that what has that structure is ultimately a vast system of interacting centres of experience of which we are high grade instances. However, all that matters here is the admission that the world of daily life is a construction whereby we cope with the reality in the midst of which we exist, but which neither *is* nor adequately *depicts*, that reality. This is not to say, however, that it may not in a certain sense depict it in terms which do a certain amount of justice to, whilst not revealing, the character which it has in literal truth.

All sorts of things are held to be true of the world we construct which reveal themselves as incoherent when we reflect deeply enough. These incoherences are rooted in the basic incoherence of the supposition that qualities, which can only really occur as qualifying components of some individual's experience, have an existence in an 'objective' world independent of anyone's consciousness. Among such qualities are shape and movement (in the full concrete form of them with which we are familiar), colour, felt texture and sound, each of which poses insuperable problems if we try really to ask ourselves just how they might occur in such a world if we take its existence as literal truth. The same is true even more obviously of aesthetic and moral qualities, and indeed all value qualities. Once this is realized, the problems found in identifying good and bad with pleasure and pain will be largely dissipated. The prime difference in the meaning of the words is that the latter point to these features only as elements of the literally existing stuff of consciousness on which the construction is based, while the former are used for them more readily in the adventures they have in the larger world of our construction.

Although the constructed world extends beyond anything we suppose anyone to experience, and has features which cannot be simultaneously realized as the contents of anyone's experience, it is more an extension of, than a substitute for, what we do directly experience, and includes this as a part, though a part some aspects of which have to be demoted to the purely personal or subjective. Thus most of the features we think of things as possessing in the inter-subjectively available public world are features which will occur as a matter of literal truth as elements of our conscious state when we perceive that part of the world. This applies both to sensible qualities and to values. The colour, shape and beauty of a building as an object in the constructed world are, when we see it as it 'really' is, qualities characterising components in the not-self side of our personal consciousness. In short, so far as things in the constructed world are actually perceived they are parts of reality as a matter of literal, not merely pragmatic, truth.

9. We can now see that the clash between our ascriptions of value and our ascriptions of pleasure and pain arises from the fact that value words like 'good' or 'bad' are used mostly to refer to values supposed present in the 'real' shared world which we construct together. It is as thus used that one learns them. Nonetheless, we are not filling the constructed world with features we never come across except as conceptual posits. Rather are we filling it with features which have a quite unconstructed presence as features of human (and perhaps animal) experience. Moreover, I have not learnt to use these words in any really pregnant way to describe situations in the constructed world, unless the constructed presence of value qualities there corresponds by and large with the way I do or would experience the relevant parts of the world

if directly confronted with them. For otherwise, either the magnetism of the values I think of as existing in the constructed world will be at loggerheads with that of things as I actually experience them or my so-called belief in them will be an empty ritual, not engaging with the real dynamics of my consciousness. Thus if my verbal obeisance to the common construction prevents my recognizing that the only values I can envisage as present in the shared common world are species of the same generic forms of the good and the bad as really qualify my most personal satisfactions and dissatisfactions, I deprive their inter-subjectively acceptable use of any personal relevance.

Thus to think of the absence of a certain book from the library as bad is to participate in, or enlarge, the construction of the shared world by imagining an objective badness as pertaining to that physical or cultural state of affairs, which is distinct from the unpleasantness of anyone's experience. We make a more striking contribution to the construction of a world including values whose existence is independent of being experienced, when we decide that some painful experience, which feels decidedly bad in the actual having of it, is 'really' a good thing. Thus if I think it good that someone suffer some unpleasant punishment, then his experience, as it figures in my imagination, is good, but the literal truth about it is that it is bad, and the only truth my judgement of its goodness can have is pragmatic. Of course, I realize that it feels bad to him; that is, indeed likely to be a main feature of the situation which I find good. But in the object world, as I am conceiving it, that badness is only an aspect of a situation which taken as a whole is good. Here the value of the situation as I envisage it in a judgement which, it would seem, can possess at most some kind of pragmatic truth, clashes with the negative value which, at least on the face of it, is all it possesses as a matter of literal truth. (I ignore the complications of a situation where the value supposed to lie in the unpleasant experience of being punished consists in some supervening sense on the offender's part of the goodness of his being thus reformed.)

It follows from these considerations that judgements of value can be assessed for two different sorts of truth. They can be assessed as contributions to the formation of the shared constructed 'real world' of daily life, in which case the main kind of truth and falsehood to be looked for will be pragmatic;

alternatively they can be assessed as identifying the presence or absence of value as something which can be there in ultimate literal truth, in which case they can only be true by pointing to a value felt in the form of pleasure or pain in some stream of consciousness.

It is, of course, no more a matter of arbitrary decision which values shall be introduced into the constructed world than it is to decide on its more 'factual' contents, meaning by these simply contents not characterised in evaluative terms, for I am certainly not denying that values are in their own way facts. Once granted the basic principles determining the way in which we construct the world upon the basis of our experience, there is little or no choice as to what it is proper to count as true by way of statement about it. (The obvious fact that we can act and change the world thereby is not in question here.) The position of the furniture in the room, the places where there is oil to be drilled for, the population of Scotland, statements about these all turn on definite facts as to what we will observe under definite conditions. So far as they report on what lies beyond our immediate experience, they only hold in that picture of the world which we base on our perceptions (perceptions which, indeed, could not be quite the same if they were not being fitted into such a picture) but, granted our type of perception and conceptualisation, we simply have to set out to discover what is the case, without having much or any choice as to what that shall be. How far there is a common construction for all humanity is arguable. The question, however, risks becoming merely verbal, for it is unclear when we should speak of one construction with variations or of different constructions. Certainly, some element of common construction seems unavoidable. [Cf. GRAYLING.]

Much the same applies to values. Granted the basic features of a construction of a world, then things simply present themselves as having certain values. However, there is evidently more variation in construction in the case of values than in matters commonly contrasted as factual. The values with which the world is presented as replete may differ considerably from society to society, or (though more in some societies than others) from individual to individual.

Despite this, an individual usually has little choice as to the values which he will posit in the world. This will be determined partly by the society in which he has grown up; partly by special features of his own personality which, however it came to be what it is, will make him experience and imagine things with a particular kind of value charge. I do not choose, for the most part, what I will find good or bad, ugly or beautiful. The things which I immediately experience have their own real values, and if I think of them as having an existence independent of my consciousness of them, I will tend to think of them as having the same values apart from the way in which they sparkle or darkle for me.

Still, even if we cannot freely choose our values, they may change as a result of fresh thinking and experience. The question naturally arises: Is there any sense in which we can regard ourselves as coming up with answers which are right or wrong? More generally, can judgements of value, when they concern values as supposedly belonging to a shared object world, have any kind of truth or falsehood?

Why should it be different with values than with the factual features of the constructed object world? Even though there is no literal truth in either case, there is, evidently, some kind of pragmatic truth, some way in which these judgements can work effectively for us or otherwise. However, although such value judgements and factual judgements are on a level as aspects of our construction of a shared world, one cannot entirely equate the sort of pragmatic truth it is appropriate to aim for in each case. The pragmatic truth of factual judgements, as contributions to the construction, is largely a matter of the extent to which they will help us predict and control experience so that we can satisfy desire or, what comes to the same, experience what we feel as good rather than bad. The point of value judgements, as contributions to the construction, will also be to promote good experience, and prevent bad, but since they are themselves judgements about the good and bad, with the potential to become active desires, they must have a more intimate relation to this goal.

Not that the pragmatic truth even of 'factual' judgements about the constructed world is entirely a matter of their ability to help us satisfy our desires. For, first, their very utility turns upon their relating appropriately to (partly by incorporating some of) the literal truth about what is presented to us in perceptual experience. In this respect their relation to the literal truth about the contents of experience is quite similar to that of value judgements. And secondly, there is reason to think that factual judgements, especially the more scientific, may give a more or less correct indication of the literal truth about the underlying structure of the 'things in themselves' which are the source of these experiences. Whether there is any way in which value judgements about things supposedly existing beyond human or animal experience can indicate any at all similar literal truth is doubtful. (Eventually I shall suggest they may sometimes give a hint of values existing at the level of the underlying reality.)

Thus within the totality of factual and value judgements by which we construct our ordinary shared world, and which would be counted true on the basis of usual sorts of test, there is much that can only be true pragmatically, but there is also a core of what can be true quite literally. Where more than ascription of structure is concerned, this core consists of judgements we make about what individuals experience. Now it is much more important in the case of value judgements, than with factual judgements, to distinguish between the core of judgements which can possess literal truth and those which, because they concern what is unobserved or unexperienced, can only have pragmatic truth. For the only basis on which it seems sensible to judge the construction is the extent to which it helps us cope with, and live satisfyingly within, reality, and that means by its effects on values whose occurrence in the form of pleasure and pain is a matter of literal truth. And on the face of it, that means values which occur in the consciousness of individuals much like ourselves. (There will be some qualification of this later.)

So whenever we wish to be guided by a sense of values which are really there, not merely created by our positing, we should concern ourselves only with values which are realized as features of the conscious experience of individuals. When we are right about what values of this sort are being realized we are apprised of literal truth such as simply does not arise with other values. It is not, indeed, any more true of values than of the ordinary sensible properties which form the character of physical objects as we ordinarily think of them, that they only exist as components of sentient experience like ours. But with these latter characteristics, it is only on rare occasions that we need recall this. With values, the occurrence of which determines the whole point of the construction of an object world, as of all human activity, we need to recall it whenever we reflect critically on the point of human behaviour, that is on matters ethical.

In saying this I am drawing a conventional distinction between facts and values in order to bring out the main point more clearly. Actually, sensible qualities as they occur in experience are themselves values, insofar as they are forms of pleasure and pain. So a more accurate way of putting things is to say that when we are deciding on our ends it becomes important in a way it does not when we are merely thinking about means, or facts of which we need to take account in pursuing whatever goals we have, to distinguish between the literal truth of what is to occur in felt experience and the pragmatic truth of what goes on in a world beyond it.

That is not to say that it is only by the thought of values actually present as elements of experience that people are or even always should be actually guided. For apart from anything else, much of the value literally realized in our lives may turn on, or even be the direct value of, the experience we have in thinking of the world as replete with values some of which exist without being qualities of any experience.

Is it merely a dogma that the test of the pragmatic truth of judgements which create the constructed world must be their contribution to experienced good, or the absence of experienced bad? Well, we must recognize, first, that for the most part, we are bound to think of them as literally true, and advocate them as such. Only on the basis of philosophical reflection does one realize they cannot be so, and that yet nonetheless there are many of them which it remains right to affirm and, indeed, in the course of ordinary life, continue to think of as literally true. (This, indeed, is the only way in which we can affirm them, for every judgement in the actual affirmation of it is – in a manner – thought of as literally true, and can only be ascribed a merely pragmatic acceptability in a separate reflective moment.) One then faces the question, how they are to be evaluated, if not for literal truth, and can hardly fail to give pride of place to their promotion of good experience, and hindrance of bad. Why so? Well, first, the best explanation of their actual prevalence will be that they have been useful, that is, have developed as a result of the fact that they assist us to survive and live lives which are the better for our holding them. If we still see reason to retain them, when we realize that they are not literally true, it must be because we still see belief in them as being or doing good in some high degree. Some of that good, indeed, may be good, or prevention of bad, which, on further reflection, is seen itself only to be there as part of the construction. After such reflection, one will only want to continue constructing that good, if it serves a further good, and this good must be a good which magnetises us (or the absence of a badness which repels us) because we conceive of it as really there, and perhaps eventually experience it ourselves. Thus in the end the good for the sake of which we favour making judgements, which we now see as themselves incapable of strictly literal truth, must be a good (or the absence of a bad) whose presence is, or can be made, a literal truth.

10. We must now say a word about the difference between moral values and others. Values, I suggest, are morally significant if it is especially important that people should be aware of them. That is, a value is morally significant to the extent that it is very bad that people should be unaware of it. The sufferings of victims of oppression are morally significant (dis)values because it is particularly bad that people should not recognize, and take account of, their badness. (There is nothing circular in this account for the badness of insensitivity to morally significant values is not specified as itself a morally significant value. There is, however, nothing to stop us going on to make a further judgement to this effect.)

A useful narrower conception of what it is for values to be morally significant, is that they are those values respect for which should be enforced, perhaps by law, but certainly by public opinion. The badness of lack of respect for them presents itself as an evil so urgently requiring correction that life should be made uncomfortable for those who lack it. We do not, indeed, merely judge it good that they should be made uncomfortable. Our thinking these values moral ones, in the narrower sense, is one with the fact that lack of respect for them evokes a peculiar hostility, presents itself in a peculiarly nasty light. But on reflection we would only say that these really were morally significant values if we judge it good that failure to respect them should evoke such hostility.

Where it is only especially important for some to be sensitive to them, they are morally significant, in relation to these people but not so for others. By the moral goodness and badness of persons is understood primarily their sensitivity or otherwise to values which are morally significant in relation to them. In our culture these are thought typically to be values realized in the consciousness of others to which absorption in our own projects is liable to blind us. (See chapter nine for more on this.)

As to what is normally meant by calling actions morally right or wrong, it would seem that these descriptions apply to actions primarily as being or not being manifestations of sensitivity to the morally significant values liable to be realized or otherwise in performing them. In deciding whether it would be wrong for me to do something I am concerned to act in such a morally sensitive way. The 'external' rightness or wrongness of an act is a matter of whether it is what a morally sensitive person would have done. The 'internal' rightness or wrongness of an act is a matter of the role actual moral sensitivity or lack of it has in its actual performance.

Thus moral values are distinguished from other values by the special value we attach to people being sensitive to them, and in particular the special degree of awfulness we see in insensitivity to them. But though moral values are thus distinguishable from other values, they are still just one species of those generic qualities of good and bad of which the various forms of pleasure and pain are the most obtrusive species. To think that there are values realized other than as felt qualities of modes or contents of experience is to join in a construction which can only be legitimated by reference to the immediate values of actual experience. For it is only such values which can beckon us or threaten us when we look at things in the light of the metaphysical truth that the only values literally present in the world are those present in consciousness. If there are values in the world beyond human and animal experience, figuring there in literal truth and not only from the point of view of a useful construction, that can only be because the world contains experience which is not that of man or animal. Even if there are such values in nature at large, as we may feel that there are when overcome by the wonders of nature, they must be distinguished from any values which only pertain to those wonders in that version of them which figures in our constructions.

11. The word 'ought' stands rather apart from other ethical words in that the sentences in which it typically figures have as much the feel of an imperative as of a statement. Certainly the affirmation that I or another (morally) ought to do something often means much the same as the statement that my doing it would be morally right, and my not doing it morally wrong. That is, it expresses – if we are correct in our account of rightness and wrongness – our vision of the unattractive insensitivity to these considerations that we would show if we acted otherwise. However, it can also be seen, rather, as an instruction which one gives to oneself or others in response to such considerations. Thus it may figure as a stimulus to action, rather in the way that the emotivist supposes that ethical language in general functions. Sometimes, indeed, when I say of myself that I ought to do something, it is not so much a *stimulus* to action, as a verbal cog in the machinery of the will, a phase in the will's mobilisation of itself, that is, of the process in which the envisagement of values flows into action. However, in any kind of moral thinking, which deserves the name of thinking, such stimulation or verbal machinery rests ultimately on an envisagement of certain possibilities of action, and what would flow from it, as having some kind of positive or negative value.

12. Our account of moral judgement is akin to G.E. Moore's inasmuch as we take ascriptions of a quality of intrinsic goodness or badness to things as basic. However, Moore was utterly opposed to any identification of these qualities with those so-called natural qualities which constitute a pleasure a pleasure or a pain a pain. To this I say that the essential magnetism and repulsiveness of pleasure and pain do indeed mark them off from what are typically thought of as the natural qualities captured in what an emotivist would call purely descriptive language and that this identification allows us to understand why moral and other value judgements are intrinsically motivating in a manner which Moore makes no attempt to explain.

Moore would insist, of course, that it can never be self contradictory to deny goodness or badness to a situation characterised only in 'natural' terms like 'pleasurable' and 'unpleasurable'. My response is that, indeed, ordinary thought allows us to dissociate judgements of value from judgements about the presence of pleasure or pain, since we may project goodness and badness, of the sort we feel in our pleasant and unpleasant experiences, on to situations where we do not think corresponding pleasure or pain is actually felt. In such cases the situation which has the pleasing or unpleasing quality for us in our imagination of it is one which we know does not have the kinds of effects which one involving pleasure and pain would do and the attractive or unattractive air it has for us may even rest upon our denying that it is attractive or unattractive as an experience to people more immediately involved. My claim, however, is that this is an incoherent way of thinking, and that, even though he did not realize it, the goodness and badness on which Moore's mind was in fact dwelling when he spoke of intrinsic value was simply that magnetic or repulsive quality which makes pleasures pleasures and pains pains, and constitutes the difference between the positive and negative in existence.

Moore's objection to hedonism is closely associated with his belief that there is an irreducible plurality of goods and evils. But once we see that pleasure and pain exist in radically different and incommensurable species we need no longer see hedonism as reducing all that is good or bad to certain uniform sensations. It is total situations which are pleasurable or unpleasurable, good or bad, each in its own specific way. Our position no more reduces all values to one than Moore's, indeed it has less tendency to do so, since Moore's formulations tend sometimes to suggest that what really matters is the presence of a single unchanging quality of goodness.

Moore could argue that we take no account of the special sort of necessity which makes a situation of a certain natural sort possess a certain intrinsic value. However, I have contended that each unique sort of pleasurableness or unpleasurableness is intrinsically related to the whole character of the experience, or component of experience, it qualifies and pertains only and necessarily to it. There is not some separable identical residue which pertains to a joyful experience and the kind of experience which could count as a depressing version of the same, though, they can, of course, have much in common.

Appendix A

The views about pleasure, desire and volition taken in this and the previous chapter are compatible with a wide range of approaches taken by psychologists to the explanation of behaviour. It is, for example, compatible at one extreme with the behaviourism of B.F. Skinner and at the other with cognitive dissonance theories and theories of intrinsic motivation of the kind espoused by such writers as E.L. Deci. They also fit in well with the 'experimental hedonism' sometimes favoured by P.T. Young. In one way or another most psychological theories of motivation operate with the notion that some states are good from the point of view of the organism, and others bad, and that motivated behaviour, as opposed to purely instinctual reaction, is behaviour which is explicable, in one way or another, by its being what the past experience of the organism indicates as likely to promote the former and prevent, limit or terminate the latter.

Psychological theories do not normally differ over the basic guidance of behaviour by experiences falling under the polarity of pleasure and unpleasure or at least by their presumed physical equivalents in the brain or nervous system. The difference is primarily as to what is rewarding, what punishing and why. Is it basically the reduction of stimulation that constitutes reward, and stimulation that constitutes punishment (or at least distress such as activates behaviour found effective in removing it)? Are reward, punishment and distress always connected with the removal or presence of tissue deficiences, in tissue beyond the central nervous system of which the state of the latter is a symptom, or does the central nervous system have specific 'needs' of its own, as, for instance, the resolution of 'dissonance', or the maintaining of a level of challenge neither too slight nor too great as in theories of intrinsic motivation like Deci's?

Psychological theories differ also as to the nature and complexity of the cognitive patterns involved in the processes by which experience guides the organism to rewarding and away from distressing degrees or patterns of stimulation, on the detailed ways in which fresh habits of behaviour are stamped in by reward and punishment, and the conditions under which the upshot is advantageous to it or goes wrong. It remains the case that most psychological theories operate with the same basic polarity of reward and punishment or distress.

So the psychological claims I have made in this and the previous chapter are little more than phenomenologically filled out statements of the presuppositions of most motivational psychology. They must be faulted, if faulted at all, as phenomenology, not as failing to square with what is empirically observable from an extrinsic point of view. The most essential claim is that all reward, at the level of consciousness, takes the form of contents of experience with a characteristic falling under a common genus which can be called either the pleasurable or the good, and all punishment or distress takes the form of contents of experience with a characteristic falling under a common genus which can be called the painful, unpleasurable or bad, and that these two genera are definite generic qualities constituting an identical common element present in all their specific forms. These specific forms differ from each other in a manner that is qualitative rather than quantitative; nonetheless in talking of the basic generic qualities of the pleasurable and the unpleasurable we are not merely gathering together a set of varied qualities which happen all to function similarly as 'positive' or 'negative' reinforcers of behaviour which has produced them. Rather are they distinctive generic qualities, identifiable introspectively apart from their effects, though there is a necessity in the way in which they tend to promote or check behaviour as it is felt within the consciousness in which they occur.

This necessary tendency holds within consciousness as a link between felt quality and felt behaviour. How it connects with what corresponds to it in the physical, – brain, nervous system, muscles, sense organs and so forth, – that is, in the 'public' world, is the heart of the whole issue of the mental and the physical. I cannot go into that here. My views are set forth in *The Vindication of Absolute Idealism*, especially chapter four. But clearly cause and effect, as they occur within an individual consciousness, somehow either spill over into, reflect, or are in some way not really altogether distinct from, cause and effect as holding between brain, body, behaviour and environment.

All in all, the psychological claims made use of in my ethical argument are little more than the views of common sense, but it is a common sense which remains as the basis, whatever the terminology used, of most psychological theory. See, for example, *Intrinsic Motivation* by E.L. Deci, *Personal Causation:*

The internal Affective Determinants of Behaviour by R. De Charms, *Science and Human Behaviour* by B.F. Skinner, *Emotion in Man and Animal* and *Motivation and Emotion*, by P.T. Young.

This must be qualified, however, by noting that academic psychologists who think in terms of positive and negative reinforcement are all too ready to approach these in a purely external way, taking these concepts and such others as 'drive' either in a basically behaviourist fashion, or as so-called intervening variables, ignoring the phenomenological dimension. Otherwise, psychology could not be the truly 'dismal science' of the twentieth century, with its grotesque record of the electric shocking of rats and the constant deception of human beings. (See *Motivation: Selected Readings*, ed. Bindra and Stewart, Pelican, London, 1966 for the justice of this comment, at least up till that date.) Setting out from an insight into what makes animal organisms 'tick', which one could hardly have without being one, enjoyment of their own investigative powers or the need to do something in their departments, has prevented the inward significance of the concepts they use reverberate in any kind of sympathetic manner in the consciousness of far too many psychological researchers. Having initially drawn on introspection and empathy to get the abstract structure of some basic truths about the springs of human and animal behaviour, psychologists have emptied the key words of their science of the moral power which springs from real imagination of their referents.

Appendix B

It will be recalled that a main argument of J.L. Mackie's against ethical objectivism was that it postulated intrinsically prescriptive or motivating features of the real world, to which McDowell replied that in this they are comparable to many other features of the only real world there is, one whose character is inseparable from human response thereto. It will be seen that I follow a middle path, treating values actually realized in experience as literally true and intrinsically motivating features of the world, and other values, rather as Mackie did, as pertaining only to a more or less beneficent construction.

CHAPTER VII

A Kind of Utilitarianism

§1. Our Conclusions Point Inevitably to Some Kind of Utilitarianism

If the discussions of the last two chapters are well taken, then a kind of utilitarianism emerges as the one fully rational form of ethics. Utilitarianism holds that value is present wherever and only where pleasure is present, or, if the value is negative, where pain is so, taking pleasure and pain in the broadest possible sense. We have seen that this is the literal truth of things. Judgements ascribing value of any sort to anything are, in fact, judgements of the presence of a quality which, properly understood, is one and the same as pleasure, or, in the negative case, as pain. Thus moral thinking which aspires to literal truth must be utilitarian in character.

We do, indeed, often ascribe value to things not conceived of as states or components of consciousness or even to experiences whose hedonic quality is in conflict with it, but this is part of an ultimately incoherent construction, which can only be true or false in a pragmatic sense. We have seen that ultimately this must turn on the good it does for us and that this good must ultimately rest on some literal good promoted in the form of pleasant experience won or unpleasant experience prevented.

But even if there is positive and negative value literally present in the world wherever and only where there is pleasant and unpleasant experience, why should each of us individually concern ourselves with values which will not be experienced by us personally or by anyone for whom we particularly care? And even if we admit that other sorts of value judgements can be true in a pragmatic sense in terms of their effects on actually felt values in the form of pleasure and pain, why should such 'truth' concern us except in so far as it affects the happiness of ourselves and those dear to us?

My answer in brief is that the good is essentially magnetic to one who adequately conceives it, and the bad repulsive. The same essential mechanism by which ideas of our own good actualise themselves in behaviour, and of the bad do the opposite, must operate with any idea of the good or the bad so far as it does real justice to its object. That the good and evil to be realized in the experience of others does not matter to me as does the good and evil to be realized in my own experience (or of those especially dear) is because I do not adequately conceive it. Thus if we could let the truth of things play upon us in any fully realized way we would accept a utilitarian ethic. To the extent that we are not utilitarians we are being irrational, in the sense that we are not facing up to how things really are, but treating value in our own experience as a reality in a way in which value elsewhere is not.

However, there are perhaps two questions here. First, there is the question why we should not acknowledge the truth of the matter but not pay the kind of attention to it which a utilitarian ethic would prescribe. Second, there is the question, why we should not avoid confrontation with the often uncomfortable truth that the experiences of others are (in general) intrinsically good and bad in just as real and intense a way as ours are, and as such demand our trouble and concern just as do those of ourselves and our dear ones.

I have indicated my answer to the first question, namely that there is no possibility of acknowledging the truth of the matter in any full way without being swayed by it. One cannot envisage the good without being drawn to it, or the bad without being repelled by it. The idea of the good has an intrinsically pleasant character which the basic nisus of our consciousness towards the pleasant will seek to intensify by making it the idea of something we can take as actual, and the idea of the bad has an unpleasant character which the basic nisus of our consciousness away from the unpleasant will seek to intensify by action which will allow us to conceive of it as something of which reality is free or being freed. What stops us being morally good is not that we fully believe in the joys and sufferings of others and do not care about them, but that we do not fully believe in them. Of course, in words we acknowledge that others feel just as fully as we, or those dear to us, do, but we do not really conceive the pleasures and pains of others, apart perhaps from a few dear ones, as genuine realities on a par with our own. But normally when we see or envisage the suffering of another our mind halts at its symptoms. Verbal registration of these as manifestations of suffering, even a particular kind of look they have for us as such manifestations, may prompt us to some slight reaction of sympathy, but stop well short of any real entering into the other's state so that we respond to it as a reality on a level with what can enter directly into our own consciousness.

I am sure that many readers will insist that although one cannot envisage a situation in any full way as involving pain for oneself without this affecting one's will negatively, this is not so with the pain of others. Similarly with the attractiveness of pleasure.

'What of the sadist?' I will be asked. And what of the person who glories in the discomfiture of his enemy, or in the punishment of one he deems wicked? May he not envisage that other's pain in a very full way, and have no impulse to prevent it?

My general answer is that in all these cases the painful experience of the other is envisaged as good, and thus not envisaged as it really is. But I doubt that it can be envisaged just as it really is, with the unrelieved badness it has for the person enduring it, and not act on the will.

It is no problem for our account that the pain of another can be imagined as something which is good. That goodness can be conceived of as pertaining to something which at another level is bad, only means that the overall value quality of a situation, as we conceive it, may contrast with some of the value qualities we think of the situation as containing. There are genuine cases where the total value of a situation conflicts with included values. In thinking the unrelieved pain of another good we foist this possibility on a case where it does not apply, since the good we think of as imposed on the bad only exists in our imagination.

In the case of sadism I suggest that the clue to the matter lies in the strong connection between sadism and masochism. The difference lies only in the divergent directions in which a common fundamental fascination with pain has developed. For the sadist pain in general, or of some particular sort, is envisaged as overlaid by pleasurable excitement, and he cannot think of the pain of the other without incorrectly supposing that it has this exciting quality. Thus he envisages the pain of another as shot through with an exciting quality, which to the extent that the victim is free of any kind of masochist delight in it himself, the actual experience entirely lacks. I do not mean that the sadist exactly thinks that the victim is enjoying it, for he is aware of the way in which the experience does not mesh in with the victim's behaviour, and thoughts, in the way that this would normally imply. Moreover, part of the fascination of the pain for him is of course the resistance to his assaults it sets up (as is, of course, also true of masochism). All the same, the exciting quality it wears for him is thought of as really pertaining to the experience. If he tried to conceive it as it is in its own stark unrelieved awfulness as phase of the victim's consciousness it would not act on the sadist's will. For what acts on the will is the envisaged character of that to which it is directed, not something which is envisaged as without any magnetic quality and to which the will turns in some essentially arbitrary fashion.

If we move away from sadism, and think rather of more ordinary sorts of callousness or brutality, it is doubtless true that these sometimes go with some envisagement of the pain caused. That this is not effective may turn partly on the mere feebleness of the imagination, partly on the fact that what is really imagined is the external expressions of pain. But some will insist that there is also the thought 'It is not I, but he, who is suffering'.

However, I believe it is built into the very nature of our separate individualities that, in thinking of an experience as someone else's rather than mine, I think of it as not having quite the same intense reality as it would if it were mine. We can recognize this in a kind of notional way, but one who could really bring home to himself the illusion involved would be ceasing to be a separate person in the ordinary way at all. To some extent this occurs when love is great, but not fully enough to destroy personal identity. This is partly because the experiences of others must be characterised in fairly general abstract terms, while our own experiences are not only felt, but also conceived, in a much more full and concrete way. If there is a special timbre, as I believe there is, to each person's consciousness, this is quite inevitable so long as the barriers of personality do not quite break down, but even without this supposition it may be granted that the experiences of others are mostly conceived only in a very general kind of way. Thus there is much in what Schopenhauer says, and later Josiah Royce, as to selfishness resting on an illusory sense of our difference from others. Adoption of a utilitarian ethic turns largely on the kind of partial correction of that illusion which we can make on the basis of abstract reasoning.

But we still have to face the second question: why not turn away from the pain (or less cruelly the pleasure) of others, if it is the case that it cannot be envisaged properly as a reality without pressing all sorts of unpleasant sacrifices upon us? Well, that is, of course, what I, like many others, mostly do and it is not clear how the question should be answered or to what it really amounts. You should not do so, in the sense that if you confronted the truth of things and let it play upon you, you would not. One can also say more strongly, you ought not to do so, in the sense that if you were more sensitive to what is in literal truth good and bad in the world you would not do so, and that the degree of your insensitivity is bad in its total upshot as a matter of literal truth. So if you have a wish to live in the light of the truth, you will open yourself to the play of these values upon you. If you do not want to live in the light of the truth, that means that the influence of your own directly experienced pleasure and pain, and of the ideas of these, and of the pleasures and pains of those dear to you, is too powerful. To realize that this is how the situation really is, will make you try to improve upon it, but very likely you do not realize it, or at least only in a verbal kind of way. That is just how the author of this book is aware of these things, for the most part. The theory of ethics is not going to make either him or the reader into saints, though it may do something to make some of us a little better than we would have been otherwise.

Not that the opposite, namely constant total sensitivity to the welfare of all others, so far as it might be known, is either a desirable or a possible state for human beings. Our predominant selfishness is an expression – as I see it – of the fundamental urge to keep themselves actualised of our own particular personalities, in virtue of which the experiences of one person are bound to be qualitatively different from those of another. Without such an urge there would not be that variety of ways of experiencing the world which is one of the best things about it. Somehow that universalistic utilitarian ethic which gives the actual truth of things must exist in harness with one of personal self fulfilment. This will be explored further in section 3.

§2. Utilitarianism must do Justice to the Value Pertaining to our Construction of a World in Which not all Values are Felt

Our view does not deny the importance, and indeed inevitability, of our sustaining the construction of a world in which values pertain to things which are not conceived as anyone's mere personal experience. It will, however, think that for critical reflection the values of the constructed world only matter to whatever extent they, or the belief in them, are values realized in immediate experience. For at our most rational, when we distinguish construction from reality, we will judge things from the point of view of the reality.

However, my use of the expression 'immediate experience' must not be misunderstood. I use it to refer to any totality the nature of which constitutes what it is like being some particular sentient individual or to some element of such a totality. Chief among the values realized in such totalities are those which pertain to things perceived in those versions of them which are what they directly are for us. This always depends to a great extent on past experience, and the culture to which one belongs. As I look out of the window, I see a part of Georgian Edinburgh with people walking along the street and cars driving along the road. These would not present themselves to me as they do if I had not been educated in the use of certain concepts, and did not have a particular way of thinking and feeling about things. Nonetheless, they are immediate presences in my consciousness and their value, being something actually felt, is a value which exists in literal truth. Thus those parts of the constructed world which are presented rather than merely represented come to swell the totality of what exists in literal truth and their value is among the most important literally existent value of which we know.

In contrast to such actually felt values stand values which are merely conceived, in particular the values supposed to belong to most of the world of common sense insofar as that is something merely believed in. Since such values are only ascribed in thought, the only real value they bring into the world is the value of thinking of them.

Yet the distinction between value which is actually felt and that which is merely conceived is not a sharp one. To whatever extent our thought has the kind of imaginative richness which allows the nature of things as we conceive it to play upon our mind with any degree of vividness the value it predicates of what lies beyond itself will be more or less fully there as part of its own character. For thought, if it is to bring us a real sense of the nature of that in which we believe, must characterise it on the basis of its own resources.

Much of the time, indeed, it is rather far from doing this. For it consists to a great extent in mechanisms of verbal behaviour which stimulate conduct which is practically effective for gaining settled ends, but does not display values in any essentially magnetic form. But all this is a matter of degree rather than of sharp divisions. Not only can words take on much of the emotional character of what they represent and thus bring it before us with some completeness, but they largely shape, and themselves belong to, the constructed world which insofar as it is actually realized in experience has become more than a mere construction.

§3. Utilitarian Thinking should not usurp the Place of Commitment to Personal Self Realization

The utilitarian ethics which is presenting itself to us as rational would seem to be universalistic and consequentialist. That is, it seems to bid each of us do whatever will produce the best effects on sentient experience at large and into the future so far as that is predictable, without giving any special weight to its effects on his own experience, or that of those especially close to him. It bids us do this in the sense that it indicates that this is what we will do so far as our activity is rationally controlled by an adequate sense of the realities of the world; egoism is seen as springing from a defect in rationality, however inevitable a defect. This is a feature of utilitarianism which has been particularly challenged in recent times. Moreover, there is a long tradition in ethical thought according to which the rational ends of conduct include (or on some views are entirely covered by) the end of self fulfilment of some kind. Can it really be correct to go to the other extreme and say that conduct is both morally unacceptable and irrational whenever we are guided by a self love which is not kept in thrall to a universalistic concern for the welfare of all sentient beings? Theories, it may be suggested, which condemn such egoism risk removing ethics from the realities of human motivation, however much they may claim to be rooted in these, as our utilitarianism purports to be.

I endorse this criticism, and believe that utilitarianism has to be combined in some fashion with recognition of the appropriateness of the personal concern each of us has to live fulfilling lives ourselves, a concern which is bound to be different, and which we should not wish to be different, from the universalistic concern which utilitarianism seems to advocate. (Concern with those specially dear to us will also have to be considered.) For it is built into the very nature of consciousness that it tries to expel bad or unpleasant experiences and to sustain and bring into being pleasant or good experiences. Saying consciousness 'tries' to do this, is only saying that it has an innate tendency to do so, in the sense that bad experience produces activity of a kind which, in the past, has served to expel it, and good experience produces activity of a kind which, in the past, has tended to sustain it.

This process, which lies below the level of will in any proper sense is, so I have suggested, the basis of will. The will of which it is the basis can include concern for the presence of the good and the absence of the bad from the world quite apart from their presence in one's own consciousness. However, if will is produced by a root tendency of consciousness to enhance its own being so far as circumstances allow, one would expect each consciousness to have a kind of especial concern with the enhancement of its own being which will never quite be duplicated in the kind of concern it has for the enhancement of other consciousnesses. There would be something odd if we based ethics upon the root tendency of each consciousness to enhance its own being, and then dismissed any special concern with the enhancement of one's own being as immoral. Universalistic concern is a derivative of personal concern and cannot be expected entirely to replace it.

Let us recall the message of Spinoza here. Spinoza takes his stand on the individual conatus of each creature, that basic urge for survival and enhancement of its own being which gives it its individuality. He claims that we are all basically directed towards our own self realization in this manner, and that it is quite futile to regard this as a matter either for repining or self congratulation. What Spinoza tries to show is that if we think it out rationally we will see that this inevitable goal of self realization will best be forwarded by a life of virtue such as he describes.

Our own approach has much in common with this. There is a basic nisus within each consciousness to enhance its own being by bringing good within it and expelling bad. Where we go somewhat beyond Spinoza is in claiming that since we have ideas of what goes on beyond our own personal consciousness, and these ideas will themselves be good or bad according as to whether what they represent is good or bad, we have an inevitable tendency to favour the promotion of the good or the reduction of the bad in the world generally, so far as we properly understand what is going on there. On this basis we cannot avoid a universalistic concern for the welfare of all, cannot avoid it, that is, if we use our reason properly to understand the world in which we live. But it would be odd if the somewhat Spinozistic considerations which have led us to see universalistic utilitarianism as the ideal commended by reason, did not also point up the propriety of taking personal self realization as a goal.

A strong element of egoism in our motivation is not some merely contingent fact. The idea of a person whose motivations were purely universalistic is almost unintelligible. If I suffer pain I struggle to relieve it, or make some special effort to put up with it. If I suffer thirst I take steps to find a drink. If a painting in a gallery attracts me I go over to have a more careful look at it. If I love a particular piece of music which haunts me, I put on the record, or, if I sing adequately or play a suitable instrument I am inclined to go on singing or playing it. This basic nisus of an individual consciousness to expel the bad from itself and sustain the good within it is at the core of what it is to be one particular conscious individual. For it seems that the very identity of a consciousness across time lies in the fact that the responses stemming from the later consciousness are conditioned by the rewards and punishments encountered by the earlier. All higher sorts of connection constitutive of personal identity seem to stem from this.

Thus my concern with the fact that others (especially remote others) are tortured, that others are hungry or thirsty, or that they have a chance to enjoy life and beauty, is bound to be less primitively urgent than this. It can be absent without my ceasing to be an individual person, as the direct concern with what is going on, or is about to go on, within my own consciousness cannot. It is bound to be some kind of derivative from the concern I have with the presence of good or bad in my own consciousness.

Degree of concern with the welfare of those one is not encountering in one's personal life varies greatly from person to person. However, considerable concern with the welfare of some few with whom one is personally involved is the norm. Certainly parents normally care in a very direct way about the welfare of their children, and fond spouses and close friends may feel for each other in the same kind of direct way. To torture someone's child may be more effective as a way of controlling him than to torture him himself. How far are we to consider this kind of concern as egoistic when we seek to determine the role, actual or desirable, of egoism in human life?

The fact of the matter is that the sight or thought of someone close to one suffering is distressing in a direct fashion and consciousness struggles to remove it by doing something about the suffering. Is one to say that one is trying to relieve oneself of an unpleasant experience, or is one to say that one is trying to relieve the suffering of the other (while granting that the fundamental mechanism of this is the tendency of consciousness to operate so as to prevent that of which the representation within itself is painful)?

In truth, such questions are on the artificial side, as we can learn from thinkers like Bradley who insist that the individual self is so tied up with its relations with others, that its own self realization passes into, and is not sharply distinguished from, its need to improve the social world within which it exists. The most one can say is that action motivated by concern with the welfare of those with whom one has close personal ties is rather nearer to the quest for self realization than is action, or holding back from action, motivated by concern with the needs of those whose lives are not closely bound up with one's own through ties of affection.

It may be objected that the primitive urge to enhance one's consciousness cannot explain the love for another which would not seek relief from the thought of their suffering by simply diverting thought away from it, although that might well be the best expedient for the removal of one's own sympathetic suffering and one which would save the trouble of providing assistance. However, one cannot set out to divert oneself from thinking of the suffering of another without passing through the picture of oneself as basely unconcerned with the other's welfare. Moreover, the very attempt to divert thought from something will continually bring one back to an idea of it from which one can only find real relief by some action of assistance.

Besides, the concern of a human being is with the world in which he conceives himself as living. Now while in theory one can distinguish elements of this world which are immediately present as an element within one's own consciousness, from what is only there by way of representation, there is in fact no sharp separation between these two. From a phenomenological point of view, at least, there is at most a difference of degree. The sadness of another person as a quality of the version of that person which figures in my world, and the sadness this represents in the separate consciousness of that other, are not things which I can distinguish in ordinary practice. Maybe even at the ontological level they are not as sharply distinct as all that. From the point of view – for example – of the metaphysics of A.N.Whitehead, the sadness of another can be thought of as actually so to speak injecting itself into my consciousness. However that may be, the world which I am primarily concerned to enhance is that which is most fully there for me, whether I would say at an extremely reflective level that it was there immediately or there only by way of a representation of it. Thus this objection overlooks the way in which the life of the other, with their joy and suffering, can be an immediate value present within my own consciousness not merely represented there as an external reality of which cold reason tells me I should take account but which concern with the quality of my own experience would bid me ignore.

Thus it is often meaningless to distinguish egoistic concern from concern for the welfare of those close to us, and in the case of those who have devoted themselves to the welfare of large masses of humanity it is often doubtless equally meaningless to attempt to sort out egoistic and altruistic motivation. It is not that the truth is difficult to disentangle but that the facts of the case do not sort themselves out in that kind of way. Perhaps the distinction has most sense as marking a contrast between one's attention to one's own most basic physical urgencies together with one's pursuit of one's own main personal projects (including the welfare of those with whom one is personally most linked), and action prompted either by one's having stood back from these in order to judge their legitimacy in the light of the neglect of the interests of others to which they may be leading one, or by the sudden vivid recognition of the needs of another or others from whose lives one gets no obvious personal enrichment.

Sometimes utilitarianism is represented as the enemy of any serious attempt to live out one's own personal life in a satisfying way, and as calling each of us to live constantly as merely some kind of contributor to an impersonal totality of sentient experience. That is hardly what any of the traditional utilitarians were really advocating, but it may seem to be the final logical thrust of utilitarian doctrine. We must agree with those critics of utilitarianism who have seen something unsatisfactory about this. But if we arrive at utilitarianism on the grounds developed in this book, we will scarcely use it to deny that it is right and inevitable that each of us should be mainly concerned to live out our own lives satisfactorily, and that it is a main task of ethics to consider how we may do this more effectively. For utilitarianism is more a development than an opponent of the basic yen for good within one's own consciousness and the kind of vindication of it offered here would make no sense to beings without that yen.

But if we are to have a theory which allows the proper claims both of self love and of benevolence, in the spirit of Bishop Butler and others, how are we to combine them? (It is convenient to use 'benevolence' to mean a concern for the lot of sentient life generally, with oneself as just one unit within this, rather than a concern *only* for others.)

One approach is this. The actually right action is that which contributes most effectively to the general good, but self love, or the concern for one's own welfare and the working out of one's own personal projects, is so inevitably strong that the only practicable aim is to encourage oneself and others to impose some limits on its ascendancy. It may also be suggested that the amount of welfare in the world is by and large going to be little if each person does not do his best (subject to acceptance of some restrictions) to promote his own, for he is usually the best judge of, and virtually always the most motivated to bother about, it.

This is too grudging in the place it gives to self love. Self love, in the relevant sense, is neither some necessary evil, nor the most effective means by which we can all add our bit to the general welfare. It is the basis of what it is to be an individual sentient being, and therefore the precondition of there being such a thing as welfare at all.

It is more satisfactory to base our ethics upon the fact that we are all concerned to enhance our personal consciousness by the presence of the good and the absence of the bad, and, in the light of this, can recognize that there is a level of impersonal and absolute truth in which such enhancement is equally desirable in whatever consciousness it occurs. Viewing things from that level we will be utilitarians, but neither at that nor any other level will we think it desirable to move away entirely from the personal viewpoint. From the impersonal viewpoint of utilitarianism we see the personal viewpoint both as its source and as needing to be retained if there is to be value in the world, while from that personal viewpoint in which we are directly concerned with our own lives we see attention to the truths of utilitarianism as something of which for various reasons we must take account. One of these reasons is that, to the extent that we are rational, we wish to live in the light of the truth about things, and this includes the truth about the equal reality of value as it exists within and beyond our personal sphere, while at a less lofty level we may see a utilitarian ethic as providing the best basis for settling on rules by which we can hope to live in peace with our neighbours.

Thus an ethic of self realization and a universalistic utilitarian one are both requirements of reason, in the sense of specifications of what one will be the more motivated to do, the more one understands both oneself and the world. (I call the concern of self love 'self realization', rather than merely 'the maximisation of happiness for oneself', to emphasise the need we typically have for some over all pattern to what we do and achieve in our lives which we can contemplate with satisfaction.)

Some balance, then, needs to be struck between these two sorts of ethic. Can we give rules for striking it? Clearly no mechanical rules are going to do justice to the richness of actual life. What can be said is, on the one hand, that a rational person will move between these two, and, on the other hand, that to see them as radically antagonistic is mistaken.

Self-realization of any fullness is impossible for one who so seals himself in his own consciousness that his will is not played upon by the needs of others, while there would be no such thing as a general happiness to be considered if people did not cherish especially their own personal projects and loved ones. Utilitarianism represents the impersonal truth in the light of which we will, if we grasp it, quite often wish to check, and on rarer occasions radically to alter, those personal projects pursuit of which is rightly the normal order of the day.*

§4. How the Relations between 'Pleasure' 'Pain' and 'Happiness' bear on the Problem of the Incommensurability of Pleasures and Pains

Let us return now to the utilitarian point of view and consider its greatest problem, that of clarifying what is meant by saying that one action does more good than another in terms of pleasurable experience gained and unpleasurable prevented. The difficulty is not mainly the practical one of determining what the truth about this is. If we can identify the general nature of the truth which would settle for us what it would be best to do, if we knew it, then we simply have to accept that it is as easy or as difficult to ascertain that truth as it turns out to be, and guess as best we may the direction in which it would point. It would, however, be a devastating objection to utilitarianism if there were reason to think that there were no such truth at all, or at least some way in which some ideas about the matter were 'more true' than others.

* Cf. NAGEL 3 on oscillation between the personal and the objective point of view. Nagel's book deals with several themes central to the present work, but I have only had a chance to read it since completing mine.

The difficulty lies rather in the incommensurability of the goodness and badness (pleasurableness and unpleasurableness) of different experiences. This precludes there being a kind of truth of the sort indicated by the ideal of hedonic calculation described by Benthamism, while Mill's qualitative utilitarianism, though on the whole an improvement, is not only vague but suggests the propriety of a once and for all grading of sorts of pleasure which does not do justice to the different needs of different times and persons.

Would it help, perhaps, if we said that our aim should be not so much the maximisation of pleasure and minimisation of pain, as the maximisation of happiness and minimisation of unhappiness? The classical utilitarian, of course, often used this formula, but defined happiness in terms of pleasure and pain. Perhaps the trouble arises largely from the inadequacy of this conception of happiness?

The utilitarian may say that happiness is 'a sum of pleasures existing in the absence of serious pain'. Or he may say that happiness is 'a favourable surplus of pleasure over pain in one's life'. Do such definitions perhaps have things the wrong way round? Should we think rather of the value of pleasures and pains as determined by the way they affect people's happiness? Evidently there is a need for some examination of the relations between happiness and unhappiness, on the one hand, and pleasure and pain on the other.

If we understand happiness as a favourable balance of pleasure over pain during a certain period of time, it would seem that happiness is not a state in which one can be at a moment. As against this, I suggest that happiness in the most basic sense is a state one can be in at a moment, not a fact about how things average out over time. We shall see, however, that although one can be happy at a moment, the experience of one moment is so infected by what has passed, and what is more or less dimly anticipated, that the happiness or otherwise of any one moment is intrinsically related to that of other moments.

If we think of happiness as something which can be actualised at a moment, we cannot regard it as only a long term favourable surplus of pleasure over pain. But we may still think of it as a sum of pleasures in the absence of serious pain at or throughout any particular time describable as a happy time. Let us explore this further.

We can say that one's state of consciousness at a particular moment is either a happy one or it is not. Crudely one could say of the state of consciousness of any moment that it is either a happy state, a mixed state, or a wretched state. There seems no real difference between saying this and saying that it is either pleasurable, mixed (perhaps in some bitter sweet way), or painful. Considered as properties of a total state of consciousness pleasure and happiness seem the same. Pleasure, in the sense in which utilitarianism conceives it as the fundamental good, is surely the very same thing as happiness, conceived of as something which can be present at one moment, while one can say equally that happiness at any one moment is simply a state of consciousness which is pleasant as a whole. And surely unhappiness and painful experience as they occur moment by moment are likewise identical.

I am, of course, presupposing a hedonistic conception of happiness, as against any of the non-hedonic notions of happiness which are sometimes canvassed. These latter ascribe a kind of goodness or badness to conscious life which is other than what is actually felt, and as such will be rejected as confused by anyone who has seen that goodness and badness have to be felt in order to be. Certainly non-hedonic notions of happiness are irelevant to the formulation of a utilitarian ethic.

So the notion of a happiness which does not take the form of pleasurable moments can be dismissed from utilitarian ethics. But for a variety of reasons it could be very misleading to infer that all that can be meant by a happy life is one in which there is a large sum of pleasures with few pains.

If one made a list of types of experience which can be called pleasures because they normally make for happiness, then it is not the case that someone will be happy just to the extent that they have experiences of this type. They may be quite insusceptible to some of these so-called pleasures. One can say that for him they are not pleasures, since they do not make for his happiness. But even if they are pleasures for him, he may not be very happy while experiencing one of them, and not necessarily because it is outweighed by anything describable as a pain, for it may be just because he is in a gloomy kind of mood. Thus happiness is certainly not a sum of pleasures, or a surplus of pleasures over pains, if that means that happiness is some kind of function of the occurrence or nonoccurrence of certain types of repeatable experience classifiable as pleasures and pains.

The way in which individual experiences are classified into types is, of course, highly flexible. Listening to music is a type of experience, listening to the opening of Beethoven's fifth symphony is a type of experience, listening to it with a musically educated ear is a type of experience and so forth. But however experiences are classified into types, it seems impossible that one should arrive at types such that one can say that the happiness of someone at any moment could theoretically be read off from the types of experience which are present in his consciousness. There are two reasons for this. First, the particular instances of the type always have some distinctive quality to them, largely determined by the context of other experience in which they occur. Second, the overall happiness is not a function of some atomic elements out of which it is composed. It may be truer to say that the overall experience has a total hedonic character of which individually discriminable elements are inseparable aspects.

Thus happiness is neither a sum of pleasures in the sense that it is determined by the occurrence or nonoccurrence of experiences of repeatable sorts, nor in the sense that it arises additively from the particular elements of total states of consciousness. It is rather a pervasive quality of experience as a whole, which is often better thought of as making its elements pleasurable or unpleasurable than as arising from some separate hedonic quality these elements possess. It is true that discriminable elements may have a hedonic quality. The beauty of the painting I am looking at is an individual hedonic quality of a distinguishable element in my consciousness. Still, what matters is whether the overall quality of my consciousness possesses that overall pleasurable or good quality which constitutes happiness.

Thus much of the discussion about the commensurability of different individual pleasures, as also of attempts to decide whether some pleasures are of higher quality than others, seems misconceived. What matters – so far as any one individual person goes – is the overall feeling tone of his consciousness in the various distinct moments of his life. His having or not having a certain sort of experience at any particular time, even his having the quite particular individual experiences he is having, matter only so far as they are contributing to making his consciousness at that very moment a happpy or unhappy one.

It is also important to bear in mind the very great extent to which what people need at a particular moment – once the needs of mere physical survival and elementary comfort have been met – is simply whatever will best release tensions of various kinds which are building up. There are various aspects of the psycho-physical unity which we are which must have some kind of exercise at regular intervals if there is not going to be unbearable tension. Sexual tension is an obvious example, and there seem to be more general needs for some kind of vigorous physical activity and for emotional excitement of various kinds. To think of the main need (once survival for as long as can reasonably be hoped has been ensured) as that for pleasure of various kinds is to forget the dominant need we have to relieve tension. It is a weakness of the feel for life manifested in much utilitarian writing that it treats life as a search for pleasure and downplays the need for this mere reduction of tension.

In general, then, what is desirable is that experiences should be available which meet the needs of the moment. These vary greatly and depend on what we have recently experienced. After climbing a hill one may want a lager. There is little point in debating whether reading *Paradise Lost* would be a greater or higher pleasure. Moreover, after reading *Paradise Lost* for three hours, one may want physical exercise, and it would be pointless to consider the comparative intensity, or the difference in quality, of the pleasure of such exercise and that of reading Milton.

If there is any point in listing and evaluating different sorts of pleasure, it is in order to evaluate the contribution they can make to a happy life, whether by reducing unpleasant tensions or by their more positive felt worth. Maybe a life entirely given over to sensual pleasure will not be a very happy one, but nor is one entirely given over to intellectual pleasures. Both are liable to be boring and lack-lustre. What matters with pleasures is that their mix and organization make for a happy life, not which of them, considered as independently conceivable and repeatable things, are best.

§5. How One's Happiness is Intrinsically Bound Up with One's Whole Life in a Community

However, what we have been saying still does not do justice to the way in which it is unsatisfactory to say that happiness can obtain in a single moment, and that consequently by a happy life can only be meant a mere succession of mostly happy or pleasurable moments.

For the suggestion that a happy life is a string of pleasurable moments, in the sense in which it is perhaps most readily taken, is indeed false. One does not attain happiness by trying to make individual moments (or any short stretches of time) as pleasurable as possible. It is a pattern of living one's life over time that makes for happiness or unhappiness. One does not best promote happiness, either for oneself or others, by concentrating on the enrichment of individual moments, but by creating long standing situations in the context of which it is good to live. Thus someone engaged in painting a picture is more likely to be happy if their concern is with creating the precise effects on canvas which they seek, than if they are trying to enrich each moment. And a social reformer who wishes to make life in his community less drab may do better to promote evening classes in which people engage in the long tough task of coming to understand something than entertainments providing a string of lovely moments.

This important truth is, however, to be understood in the context of the equally important truth that long term situations are only of value because the moment by moment experiences of involvement with them are pleasurable. They must be actualised in moments of experience to be there for anyone at all in any way which matters.

It is not merely a contingent truth that experiences, as they occur moment by moment, will only be good if they belong to some appropriate over mastering pattern of life. For the way in which each moment owes its character to the whole evolving series of experiences to which it belongs is a quite essential feature of consciousness. If you read a novel, and enjoy it, the enjoyment must come moment by moment, but you could not have each momentary experience, once you have got started, unless the previous moments were as they had been, and unless future moments were, in general terms, anticipated. Each moment experiences itself as a phase in an ongoing process, and as such could not be what it is unless that ongoing process were really and truly unfolding itself.

Thus one's life as a whole had to have its precise unfolding pattern for its moment by moment states of consciousness to be happy or unhappy in the precise way in which they were. Nor are individual moments of experience only related in this necessary way to earlier experiences of one's own; they also stand in necessary relations to the experiences of others. In a conversation it is an essential element of the character of one's experience that it is the response to certain partly grasped actual thoughts of another. And there are less immediate necessary connections between one's experiences and those of others who lived in the past. I could not have had quite the thoughts I am having now unless there had been past thinkers, such as Santayana and Husserl, who had the thoughts they had. (Both philosophers wrote pregnantly on the temporal aspect of experience.)

One's experience is similarly bound up in a necessary way with those of others when participating in any shared project. That explains why Moore's method of isolation is so unsatisfactory. A good human life is good in itself, as what it intrinsically is, but it is quite impossible that it should be what it intrinsically is and thus be good in itself in that way without being significantly related to a larger social context which includes other lives which are similarly good in themselves.

Philosophers sometimes fantasise that one might be, for all one's present experience proves, a brain in a vat of some scientist of the twenty-first century producing delusory twentieth-century experiences by electric stimulation. But he could not feed me with the thoughts present in the conversation I seem to be having with a friend, unless he had those thoughts to feed me with, or with the music of Beethoven, as apparently recorded on a tape, without him or someone else having had the experience of creating that music. So one's experiences of contact with others cannot be radically delusory. This means that the moment by moment experiences, in which value of the kind which really matters, as having genuine reality, must be realized, have to be experiences of those who are living in certain sorts of total situations. Thus a happy life is, indeed, not a surplus of pleasurable moments, conceived as realities which merely pile up without being deeply connected with each other. Rather it is a matter of the actualisation of whole situations in a whole lot of distinct moments of intrinsically connected experience each with their own particular perspective on the whole. The value must be present in the parts to exist, but it is in a sense the value of something which goes beyond the parts.

Of course, typical empiricist philosophers will deny that the relations of which I have spoken in the last paragraph can be strictly necessary. I think they are badly wrong, but cannot argue the matter further here. Provided it is admitted that the relations are deeply rooted in pervasive fact, their main ethical significance can be acknowledged.

We may conclude that the utilitarian goal should be to promote total situations such that the living in the presence of them is good as realized moment by moment in intrinsically related experiences. There is no need to look for ways of measuring the hedonic intensity, or the qualitative level, of different 'pleasures' and 'pains', whether considered as types of experience, or as individual sensations, thoughts, or feelings. What has to be considered is the role they play in making the lives of those who experience them happy or unhappy.

§6. Epicurus and the Doctrine of the Limit

If we look back to one of the first hedonistic systems of ethics, that of Epicurus, we will find another consideration which somewhat reduces the problem of weighting. Epicurus advocated a doctrine of what he called 'the limit'. According to this, once one is free of trouble, one has reached the maximum possible, at that time, of happiness. There is no need to strive for more, because more cannot be obtained [EPICURUS p. 60 §III].

This can be understood as a rather gloomy doctrine about the limited possibilities of human happiness. But it can be taken in a more cheerful way, in which, if not quite true, it is not so far from being true. On this interpretation it tells us that the removal of trouble leaves consciousness free to gravitate to a worthwhile level of happiness, since there cannot be a neutral state between that and some kind of distress.

Even if this is rather a simplification of the facts, it has something to it. We can say, at least, that life divides mainly into three different levels and that insofar as people are freed from sources of trouble, they will move to the first, and that this is so much better than the other two, that there is little need to worry about gradations within it. These may be described as: (1) a state of pure happiness; (2) a state of mixed happiness and unhappiness; (3) a state of wretchedness. The distinctions between these are much more important than any distinction between different degrees of the first. If that is so, we can see the utilitarian task as that of removing sources of trouble (including all sorts of psychological blocks which stand in the way of people engaging happily in activities which others enjoy) so that life exists so far as possible at the first level. Thereby we largely dispense with the problem of weighing types of pleasure (though not perhaps types of misery) one against another; they need be compared one with another solely as resources against misery. Choosing which miseries most call for relief is less easy, but we may look for lifestyles which reduce them all.

This is not the doctrine of negative utilitarianism, according to which the sole moral aim is to remove distress. That has the absurd consequence that the best thing to do would be to exterminate all life in which there is any distress at all. If much of the point of relieving suffering is to render life happy we should hardly do so by wiping people out.

§7. Happiness, Pleasure and Pain

Thus if we take the utilitarian aim, as seems appropriate, as being happy or pleasurable consciousness, rather than some surplus of individuated pleasures over pains, the problems of weighting are much reduced. Yet it may strike the reader that this view of the utilitarian aim does not quite square with the reasons which have led us towards a utilitarian ethic in this work. This would be a mistake, but an understandable one.

Our view has been that what solicits activity, and presents itself as good to reflection, is not some kind of abstraction called happiness but a whole number of distinct forms of the good or pleasurable. Similarly it is individual forms of the bad that we avoid, not merely unhappiness in general. Surely we betray this sense of the plurality of goods and evils, if we bring everything down to the single aim of promoting trouble free consciousness.

There are two questions here. Is it at total states of consciousness that we aim or at individual realities which can present themselves in consciousness? And is the goal merely trouble free consciousness, or specific forms of the good?

(1) It is true that what typically solicits our desire are specific realities which can display themselves in consciousness, not total states of consciousness. Indeed, we often aim at producing results which we do not recognize as having their only possible home in consciousness. But the truth is that the good or the bad (in all their immense variety of forms) can only occur in, or as a state of, consciousness, since goodness and badness are one with pleasurableness and unpleasurableness. Such is the basic argument for our utilitarianism. It follows that the good can only be brought about by producing states of consciousness which either are intrinsically good or which contain what is so. And in the end there is no real distinction between these two. The total state of consciousness must either acquire its overall goodness or badness from that of its components or impose some holistic character of its own upon them. Our explicit concern may sometimes be more with whole states of consciousness, sometimes more with some possible object or component of consciousness. But these are simply specifications of the same essential goal on the basis of different sorts of abstraction from the total situations required for realization of the good. And the same applies, of course, to the bad.

(2) The answer to the second question is similarly to be found in the contrast between a more abstract and a more concrete concern. Actual value consists not in abstractions such as happiness, unhappiness, pleasure and pain, but in innumerable specific forms of good and bad whose home can only be in consciousness. Utilitarianism bids us be appropriately sensitive to the whole wealth of these. Its own formulation is, of necessity, abstract, but in recommending such a generalised goal, as that of trouble free consciousness, it does so knowing that that will be the home of innumerable specific forms of the good, each good in its own quite specific way. As individuals pursuing our own self realization we will be concerned to encounter specific goods, not

merely to have trouble free consciousness. As utilitarians we will be concerned to eliminate those sources of trouble which militate against individuals gravitating to their own favoured forms of the good.

Thus when we deal in the abstract formulas of utilitarianism we are, indeed, moving away from the richness of the diverse forms of value to an abstraction, and it is important that we do not lose ourselves there. However, it is an abstraction the point of which is to draw our attention to the concrete conditions of all types of value, and to provide a generalised guide to conduct to which we must sometimes move, if our lives are not to be unduly unbalanced in the degrees of concern we show for different values. Utilitarianism rests on a recognition that the notion of a good or evil existing divorced from consciousness is an ultimately incoherent abstraction. It stresses that what matters is the kind of consciousness which occurs, not so much because it is only concerned with consciousness as such, as because it recognizes that it is the only possible locus of the good.

Even if the discussions of the last three sections have not entirely resolved the problem of weighting, a utilitarian ethic remains a significant guide to conduct. For a great deal has been gained by establishing that the considerations which count in favour of doing something must consist in happiness promoted or unhappiness prevented, and conversely with the considerations that count against it. Even if no agreed way of weighting these considerations could be found, and even if there were no objective truth about it, so that the best for which we could hope would be well informed fundamental decisions in attitude, in which we could not be expected always to agree, it would not be an empty achievement to establish the nature of that which favours and that which counts against an action's acceptability, and to steer us away from considerations which do not ultimately come down to these.

When considering the problem of weighting it should also be remembered that moral philosophy has a double aim. On the one hand it is concerned to discover what rational backing there may be to as much of common morality as it does not seem fitting to discard. On the other hand it is concerned to improve the decision procedures available to us for coping with problems where common morality gives no clear lead, and possibly to give reasons for discarding parts of common morality. It is a mistake to concentrate all the energies of moral philosophy on just one of these.

So far as the first goal goes, it would seem that a very large part of common morality can be supported simply by reference to good effects which bring little by way of evil with them at all. It is to a great extent a guide as to how to get on with others in a way which is good for all parties. As such, we can see that it has a utilitarian rationale requiring no definite weighting of a diversity of goods and evils. And when the considerations urged in previous sections are taken into account, utilitarianism gives clear enough guidance on many of the matters on which appeal to a thus justified common morality seems insufficient.

Certainly, someone really concerned to promote the good and prevent the bad, will not regard a careful weighting of goods and evils as his first priority. This will be rather to look for patterns of action which promote happiness all round. What stands in the way of solving most great human dilemmas is not the difficulty of weighing up different goods and evils against each other, but total blindness to some of the main inherent values or disvalues involved. The problems of Northern Ireland arise not so much because there is a difficulty in weighing up precisely the effects of various policies on the interests and passionate aspirations of the two main groups, but because those who dominate the thinking of each group seem incapable of entering at all into the values which figure in the consciousness of the other. If each side could really imaginatively participate in the ideals of the other, and see that each is good in its own way, then energies could concentrate themselves on finding a *via media* with something in it for each side. The problem is that neither party sees anything in the values of the other, not that they cannot make a nice calculation as to their comparative weight.

§8. The Hedonistic Calculus Revisited

Yet it must be admitted that there remain ethical questions where the problem of weighting cannot be dismissed. This is particularly true when we have to balance the requirements of one culture, or perhaps one species, against another. Even if we cannot provide a formula for carrying out this weighting, it is important to know whether there is any truth in the offing at all to be the object of our surmises.

The root question is this. Suppose that we somehow knew the totality of the ways in which the world would be different according as to which of two actions were done, then what should make us regard one as the better totality to choose? We know that it would be the pleasure in each which would count in its favour and the pain against, for these are the only genuine goods and evils which a mind which saw into the truth of things would recognize and which would act on its will. But with each of these totalities likely to be a mix of pleasures and pains which solicit activity to promote or prevent them, how would these solicitations and repulsions combine to make one totality solicit or repel such a mind the more?

The Benthamite would have us calculate the value of each pain and pleasure in terms of its intensity times its duration, and subtract the one from the other. We have seen the doubtful sense of this. As for the Millian alternative, its main virtue is the way in which it points up the defect of the Benthamite and it offers no real solution.

Let us see if there is any alternative which will avoid the defects of each. We are not seeking to identify a practical decision method, but with forming some idea of the nature of the truth at which we must surmise as best we can when trying to decide which of two actions would have the better total set of results.

Let us begin by thinking of a consciousness filled with two pleasant or unpleasant feelings which prompt to incompatible actions. Each prompts to the action it does, because the consciousness has registered, either as a result of past experience or by innate endowment, that this sort of experience requires certain sorts of action to sustain or enhance, or if unpleasant to remove or reduce, it. What is in question here is the elementary tendency of consciousness as it were to strive for pleasure and away from pain, not conceptual knowledge of facts.

Since the actions are incompatible the organism cannot do both. Unless it is Buridan's ass it will, therefore, do just one of them. (Of course, various sorts of compromise are usually possible, in fact, and in any case the one not done immediately may be done later, but I will have to simplify all this for present purposes.) Let us say that the feeling whose promptings thus prove more powerful has exhibited more hedonic power.

If our earlier claim is correct that the causal efficacy of pleasure and pain is a necessary consequence of their very natures, it seems reasonable to conclude that the experience with greater hedonic power actually contrasts in its hedonic quality with the other in a way which can be described as being hedonically preferable. Its greater power arises from the fact that each is the kind of experience it is, and is not merely some kind of brute fact, or one with an explanation not implied in what the experiences intrinsically are.

If two pleasures or pains occur at different times, even though in a single mind, there can be no such direct struggle for influence between them. However, so long as they lie in the future, then to whatever extent they are anticipated with any degree of adequacy a kind of indirect struggle can go on between them via the representation of them. Each, so far as it is adequately represented, will solicit activity which will either bring it about, if a pleasure, or prevent it, if a pain. Suppose now that the actions each solicits, via an adequate representation of it, are incompatible. Then, provided again that we are not dealing with Buridan's ass, one of these actions will be done, and the other left, for now, undone. We will have to simplify by supposing that the degree of probability of each occurring (or not occurring, in the case of a pain) presents itself as equal, and make other such similar simplifying assumptions about the speed with which relevant action presents itself as required by their solicitations. We can then say that that experience is hedonically the more powerful in representation which wins over the other, via its representation, in soliciting the action it requires.

Of course, representations of an experience can never be completely adequate, since if they became perfect icons they would simply be those experiences. Thus one anticipated experience might prove the more powerful, as represented, rather because it was better represented than for any reason more intrinsic to it. Let us set this aside by concerning ourselves with what might be called the respective hedonic power of experiences on adequate representation. To say that one is more powerful in this way will mean that increasingly adequate representation of each would converge on the production of behaviour solicited rather by it than by the other.

Just as the greater hedonic power of one of two experiences which occur at once arises, so I have suggested, from an intrinsic contrast between their characters which can be called its greater hedonic preferability, so it seems will the greater hedonic power of one of two experiences, on adequate representation, turn upon a contrast in their characters which can again be called hedonic preferability. Since the representations owe their character to that of which they are the representations, their contrast must reflect a contrast in these.

I now turn to the more problematic case of the contrast between what Bentham called two different 'lots' of pleasure and pain, that is, to a set of results which an action might produce consisting of a variety of different hedonically charged experiences not all felt at once. Let us begin with the case where they are all to be felt by one person.

It is evident that one often chooses between different lots of experience for oneself. In this case one forms some kind of complex representation of each lot which somehow synthesises into one unit what will be felt dispersedly at different times. Although in practice this happens only in a very fragmentary kind of way it is a recognizable process in consciousness, and it is evident that such representations can be more or less adequate, and can enter into conflict with each other in their influence upon behaviour. Thus we can speak of one lot of pleasure being more hedonically powerful, on adequate representation, than another, in the sense that more and more adequate representations of each will tend more and more to favour it as the prime influence on behaviour.

Can we say that this reflects a character which belongs to each lot of feelings describable as its hedonic preferability? It is hard to avoid this, without making choice seem largely pointless. I shall say a little more shortly about the nature of this hedonic preferability.

The representations of these different lots of pleasure and pain synthesise them into one unit, and these synthetic representations struggle with each other with a result reflecting their own hedonic preferability. What determines which will be the more preferable in this sense? If it were some kind of arithmetical result of a hedonic charge belonging to each element representing a distinct anticipated feeling, then we would have basically the Benthamite calculus. But this does not correspond to how we would actually choose. One cannot say that sufficient repetition (even if one had a drug to escape boredom) of any pleasant sensation will attract as much as some great bliss. The whole thing is much more fluid and complex than this. What seems to be the case is that certain experiences expected to occur at different times are such that their synthetic representation acquires a hedonic charge and power which cannot be regarded as the sum of the charges belonging to separable representations of its components. Thus a certain variety of experience in one's life to come may have its own appeal turning on the inevitable result of its synthesised representation, and not on some calculation which could be arrived at by separate representation of each separate experience.

I turn now to 'lots' of pleasure and pain where these will not all be the experiences of one person. Although this seems very different, it really brings in no difference of principle. If there could be a mind which could know as much about, and as adequately envisage, the experiences of others as its own they would affect it as much. In practice we cannot properly enter into the experiences of others, both because we do not have the past experience needed to imagine them, and because it is a large part of what it is to be separate persons that we do not think of the experiences of others as real in quite the way in which our own are. Moreover, mere representations of our own future experience are continuous with and constantly reinforced by the steady influence on behaviour of experiences which are not primarily representations of other experiences, and it would be neither possible nor desirable that these should have no influence beyond what they would have if they were merely represented. I am not pretending that we can live in the light of the reality of the feelings of others, present and to come, as we do in the light of our own. But that is how we would live if we were living in the light of the utilitarian truth which is our present concern. Living in that light, and with awareness of the full truth of things, the representations of various lots of pleasure and pain would synthesise the scattered complex of feelings which would result from each alternative action in just the same way as they would the scattered feelings of one person, and these representations would acquire a force which would stem in an arithmetically unpredictable way from the nature of the feelings represented.

So we can perhaps equate the truth that one lot of pleasures and pains would be better than another with its nature being such that its synthetic representation would have more power, in the event of their requiring incompatible actions, on a mind with adequate synthetic representations of each, or rather that this is the result on which struggle between increasingly adequate representations would converge.

But surely it cannot be a fact about mind as such that the synthetic representations of such lots which it would form would be ordered in their effects on behaviour in a certain way. Surely even super informed and imaginative minds might vary in their preferences.

Even if this is so, it may not be as much of a drawback to this conception as it at first seems. For each of us could still surmise how we would respond if we had the knowledge and empathic sensibility of such a mind. Perhaps we will never strictly know the answer to this, but there may be reasonable indications of it. The possibility that different people would not always converge on the same response would still leave a truth about the matter which was the relevant truth for each of us. We could each know that in deciding what to do we would be trying to decide which mixture of pleasures and pains would attract or repel us more if we knew about, and could enter imaginatively, into all of them. Thus for each of us there would be a definite truth at which we were trying to guess, even if not the same as the truth which would be relevant for another. Of course, it would not be very satisfactory if the truth were always strikingly contrary for each of us, but that seems unlikely granted that it is always the same basic factors which count for or against experiences, so that the difference will only arise when subtle matters of weighting are crucial.

However, it may be unduly pessimistic to suppose that there is no single truth here. To whatever extent the principle of same cause, same effect holds, it would appear that convergence on adequate representation of the same facts about feelings of pleasures and pains to be produced would have the same effect, if the distortions of conception which constitute our separateness as persons were removed. For the more adequate the representations the more they will turn on the nature of the scatter of feelings of which they are the synthetic representation, and therefore the more the same in nature and effects.

There are perhaps three different ways in which such representations might act on different people differently.

(1) In practice our actions are never the result only of what occurs in our consciousness. Our consciousness is problematically related to an immensely complex brain system which includes far more than whatever somehow corresponds to it. This buzz of brain activity is both the medium and the distorter of the effects which our conscious thoughts and feelings would achieve if they influenced our behaviour pure. However, the direct influence of ethical thought must be on persons qua consciousnesses and the fact that the different brain buzz of different people would always muddy the actual influence of representations of pleasure and pain does not show that in themselves they do not act uniformly in whatever mind they occur.

(2) It may be said that different sorts of personality will use synthetic representations of such lots of pleasure and pain differently. Someone with a taste for variety and excitement will be attracted by quite different complexes of feeling than one who dreads pain above all. Even when they take up an impersonal stance on things the one will favour a rich mixture of pleasure for the universe at large, even at the cost of a fair amount of pain, while the other, as he universalises his sympathies, will be concerned above all to reduce pain.

But it seems wrong to distinguish personality from the sorts of feeling and of representation which tend to occur in a mind. If we think of minds whose contents are all converging on adequate representations of alternative lots of pleasure and pain they will inevitably be acquiring the single personality of a universalistic utilitarian. Certainly it would not be desirable for us all to be like this. For there to be any life at all feelings have to be at work which do not merely represent other feelings, and for there to be separate persons there has to be a limit on the goods and evils ideas of which come to finite minds. But we are concerned with the truth which would reveal itself to a mind which is itself temporarily out of the general fray and would only act for the good of others (since he would have no good but theirs).

(3) The most powerful objection to the idea that we would all

converge on the same preferences if we looked at things in a detached impersonal way on the basis of full knowledge is that there are bound to be different ways in which the representation of a mass of scattered feelings could synthesise them.

We may recall, however, that individual experiences could not have been just what they were, or so I have claimed, if they had not occurred in the context of other experiences in which they actually did occur. My feelings now are intrinsically related to what I experienced in the past, and to those with whom I interact, and could not have been identically the same in someone with a different past or different companions. Certainly one's personal present state of consciousness has its very own character considered in itself, but that character is bound up essentially with the world in which one lives and points essentially to other experiences which flank it.

If that is so, then experiences come in connected systems, so that the synthetic representation required of a particular complex of individual units of experience is adumbrated more or less fully within these units and an adequate synthetic representation could not pattern them in arbitrary ways. They themselves determine the *gestalten* into which they would be bound to fall for one who could adequately imagine them all. He would have to unite them so that the representation of each experience places it in an overall pattern in which experiences are flanked by others in the way suggested by their own natures. Thus one could not form a synthetic representation of distinct experiences which somehow placed them in the wrong time order, since their position in a temporal series enters into their own intrinsic constitution, nor one which presented them as those of someone existing in a quite different social milieu from that in which they in fact occurred (or as existing in splendid isolation in the way conventional empiricists think possible).

We can now see how a complex of experiences not all felt together can still have a hedonic preferability of its own which determines the hedonic preferability and power of an adequate representation of it. It does so because the distinct experiences are so bound up with each other that there is a kind of overall and pervasive hedonic character which is felt to different degrees in all its elements just because they are elements within that complex. To take a simple illustration, the experience of hearing a familiar symphony is not just as series of distinct experiences each with its own independent character. Rather each experience feels its own place in the series and both helps constitute and acquires its own nature from the character of the series as a whole.

The conclusion of this complicated discussion is that different possible results in terms of pleasure and pain are better or worse than each other in virtue of the power to influence behaviour to promote or prevent them on which increasingly adequate representations of them would converge in a mind subject to no other influence. Although we are restricted to highly indirect surmises as to which of two actions will have the best results in this sense, they can be backed by the whole array of reason's standard resources, and they have as much likelihood of producing convergence in this as in most other fields.

There is a quite different problem about these totalities of pleasure and pain, however, which requires mention. All of them, after all, except one will remain for ever as mere possibilities. As such it is most doubtful whether they can ever really have a quite definite character. For it is doubtful whether there is ever a definite truth about precisely what would have happened if someone had acted differently from how he did, or if there is, it is a truth only of a general kind (for example that pain would have occurred if someone had been burnt) not a truth specifying exact consequences of alternative actions (for example the precise quality of that pain). This poses a grave problem for the claim for an absolute truth in ethics. Yet no practical thinking can get under way at all, unless we assume that we can make some reasonable sense, on occasion, of the question what would have happened if we had acted differently from that in which we did and suppose that (at least when we specify the details of what we might have done instead with sufficient exactitude) it has a true answer. If the attempt to find an objective truth breaks down here it is at a point where all practical thinking involves a kind of illusion that there are alternative ways of acting which would each have its own definite results. (Perhaps it is enough that there should be definite probabilities; however, this only brings in further complications.)

§9. Three Problems

To illustrate some of these points I close this chapter with a brief indication of their bearings on some more concrete moral issues.

A question which is thought to pose great problems for utilitarian reasoning is that of the ideal size of populations. Is it good to aim at increase of human life, even at the cost of a reduction of average happiness, if it brings more to the feast of life, as the Pope once put it?

There is the obvious point that it may be more the famine of life to which the more are being invited. The whole question tends to the merely academic inasmuch as almost any form of utilitarianism will come out in favour of strong, though of course, humane measures for reducing population.

Moreover, even if there should be as many humans as possible to enjoy the gift of life, it certainly does not follow that there should be as many of them as possible at any one time. If there is a prospect of human life going on indefinitely, we can treat the number of eventual human beings as in effect infinite, so that we do not increase the number of them by having a larger population now. Whatever the theoretical inadequacies of average utilitarianism, since, if the human race does not destroy itself, we have to do with a practically infinite number of humans it is something more like average than total happiness which must be sought. Thus size of population at one time is to be determined by considering what size of the population of contemporaries will give the best life on the average for the infinite number of people who are going to exist. Obviously too large a population puts too much of strain on resources, but perhaps too small could limit the richness of life for each and all.

I suspect that such fairly humdrum considerations may be enough to resolve the standard dilemmas on size of population which have had such an innings in contemporary moral philosophy. If not, we may simply have to surmise how one would be drawn to the alternative totalities of value in human life as a whole if one could imagine them.

Of all objections to utilitarianism the most serious is that which denies its capacity to deal convincingly with matters of life and death. When it is a question of whether a foetus has the right to life, or an embryo the right to a respect incompatible with making it an object for experiment, putting the question solely in terms of sufferings and satisfactions seems to many inadequate. Indeed, it is sometimes thought that utilitarianism cannot even explain why it is wrong (if indeed it is wrong) to kill someone whose life we think more wretched than happy, if we can do so without excessive trouble either for ourselves or others.

In considering this, we must bear in mind the distinction between values which are only conceived elements of a social construction and those which, whether as the result of such a construction or otherwise, are actually felt and not merely ascribed. Wrongness, considered as an actual *quality* of actions, is primarily a badness of the former kind, not a badness really there as a feature of anyone's experience. However, it is a pervasive feature of our experience that we feel ourselves to be living in a world in which these values are present, and our life would lose its distinctively human value if we did not. The alleged fact that certain activities are bad, where the badness is not actually a quality of anyone's feeling, is not a vulgar fiction but something posited in a way of life. Relevant criticism must concern the felt value of such a way of life.

So the proper attitude to embryos and foetuses, is to be determined not only by considering the effect on their feelings, if they have them, but by considering the felt value of the way of life in which we experience them as having a certain sort of value. Thus we must ask whether investing them with the value with which we invest humans once they have been born (and which it would certainly vastly impoverish our lives not to do) makes for a life world which it is literally better to live in than any alternative to which we might move. To pose that question here is not to answer it. If, however, we do answer it positively, we should be careful not to raise the question again too often, for thereby we would cease to live that form of life in which we have seen such value.

It is to considerations of the same kind that we must turn when we consider the issue as between rigoristic and non-rigoristic forms of utilitarianism. For how far we should live with our minds perpetually open to the impersonal best can only be settled by deciding in which constructed world of values and obligations it is best for humankind, or our part of it, to live. It is always best to do the best, but it is not necessarily always best to be thinking about what is best. It may be best – and in that sense what we ought to do – to be especially open to some sorts of value and to put special emphasis upon these by invoking a notion of 'ought' here. In the end what one ought to do is surely that which one would do so far as one lived in the light of recognition of all values of which one could make oneself apprised. But the general pattern of life which one ought to adopt, in this sense of ought, may be one in which one is not always concerned with what one ought to do. This truth is complicated further by the fact that we can hardly exist as persons, and there can hardly be personal values in the world, without each of us being especially involved in the values which belong to his own more personal sphere of action.

CHAPTER VIII

Justice, Rights and Ought

1. According to the essentially utilitarian view, which we have come to see as describing the nature of the impersonal truth as to what it is best should be done in any particular situation, the basic considerations in favour of or against an action can only consist in pleasurable or unpleasurable feeling. One action is better than another, on balance, if increasingly adequate representations of the results of each, in a mind drawn only by its sense of the pleasurable and unpleasurable feelings to which they would lead, would converge on a preference for it. Since our knowledge and powers of sympathetic imagination are in practice so limited we must make use of broad principles concerning the conditions of joy and misery, central to which is the principle that lives free from serious distress will find a level of positive joy as high as it is mostly practicable to pursue.

Thought and action from the point of view of the impersonal truth, however, should alternate with thought and action from the point of view of engagement in our personal projects. The desirability of this is something which we can recognize from both the impersonal and the personal point of view. The right balance can hardly be captured in a formula. The problem is most acute when personal projects conflict and the best solution from both points of view is often then the same, one which will reconcile them, rather than favour either just as it stands.

This approach to ethics is likely to be challenged as incapable of giving any proper place to various concepts which many think essential elements in the foundations of ethics. Such concepts include those of *rights*, of *justice*, of *duty*, and of *moral goodness* and *badness*.

2. Such a challenge might come from various quarters. It might be based, for example, upon the elaborate, powerful and influential theory of justice developed by John Rawls in his *A Theory of Justice*. Rawls contends that a just organization of society and set of rules would be that on which self interested individuals would agree if they were deciding what these should be from behind a veil of ignorance which hid from them the position they would occupy in society and the kinds of human being they themselves were going to be, in terms of intelligence, skills, temperament and so forth. That is, we are to think of a proper system of justice as what we would choose as pure self interested rational agents existing in some kind of sphere prior to our obtaining flesh and blood reality.

Rawls offers solutions on this basis to such questions as how far it is right to base the happiness of some on the unhappiness of others. In particular, he recommends that inequality is only legitimate insofar as it means that the worst off members of society are better off than the worst off members of society would be if such inequality did not obtain. One of the main things he has against utilitarianism is that it would seem to favour certain total upshots in which the surplus of good over bad experience is increased at the cost of what his system exhibits as injustice to the worst off members of society. If we consider how, behind the veil of ignorance, we would refrain from what might end up in ourselves being such worst off people, in a society which is otherwise richly happy at their cost, we will see the inadequacy of the utilitarian approach.

I believe that Rawls runs together two different approaches to morality in a way which is not really satisfactory, that from the point of view of personal projects and that from the point of view of impersonal truth.

There is a traditional contract view of ethics, according to which the set of rules we should live by represents a kind of bargain we strike with others because it is in the personal interests of all of us. (Hobbes's approach was somewhat of this sort.) Socio-biological theories of the genetic basis of ethics sometimes depict natural selection as having operated in favour of those capable of entering into such bargains. (See DAWKINS.) The main ground we have for sticking by the rules is to stop weakening the commitment of others to the bargain by betraying it ourselves. (Sometimes there is appeal to a more basically moral reason for sticking by a bargain, which is upon the whole to our advantage even in individual cases where it is not. However, this is to admit the existence of a morality more basic than that brought into being by the bargain.) Although I cannot see this as the foundation of morality, it describes a perfectly respectable supporting consideration in favour of much that is known as good behaviour. Rawls sees the principles of justice in somewhat this light, that is as a kind of bargain, but it is a hypothetical bargain made in conditions we could never be in. Why we should be influenced by considerations of the bargain we would strike in impossible conditions is not clear. The only reason I can see is that it is a pointer to what is best from the point of view of that impersonal truth to which I move when I try to see things in abstraction from involvement in my own personal projects. The virtue of the bargain theory is that it supplies us with a personal motivation for acting rightly. Rawls's theory does not. By this standard our theory does better, for what it recommends as right is that to which I would be drawn if I saw things as they really are. That, admittedly, is a call to live at a level which we cannot constantly sustain in practice, and which it would not even be for the best if we did. But the truth to which it invites us to respond is one with an intrinsically motivating force. Still, Rawls is right that the view of justice which one can derive from strictly classical utilitarianism is unsatisfactory.

Our utilitarian point of view might be challenged also by philosophers who put rights (that is, moral, not legal rights) at the centre of the moral stage. (For a wide variety of views on rights see WALDRON.) One such philosopher is Ronald Dworkin [DWORKIN]. There has also been an interesting debate on rights and utilitarianism in connection with the issue of our duties to animals, with Peter Singer [SINGER] taking a utilitarian stand, and Tom Regan [REGAN] taking a stand on rights.

Those who believe that respect for rights is required in a form which cannot be brought under utilitarian thinking sometimes see rights as representing a kind of boundary round the personal welfare of each of us, limiting the encroachment on it which is justifiable in order to increase welfare elsewhere. They believe that there are certain attacks on one individual's welfare which are not justifiable even if they represent the only possible means of a greater increase of welfare elsewhere and that to talk of a person's rights is to talk of these limitations. Utilitarianism, it is said, would have to admit that if the torture of one person was the only possible means to prevent the worse suffering of others, then it would be morally desirable to inflict it. True, utilitarians may have reasons for not favouring the more bizarre applications of such an idea (as, for example, in justifying the Roman circus) on such grounds as that they would be incompatible with prolonged social peace or with fostering the kind of emotions most favourable to the general good. Still, it may seem to leave more of an opening for such ideas than is acceptable to most people.

Whether such rights are absolute or not may be disputed. Even if they are not absolute, they represent a barrier to the use of one individual, or set of individuals, for the sake of others, which it requires something more than a mere increase in the surplus of happiness over unhappiness to justify. Tom Regan extends rights of this sort to animals, who have a right not to be the victims of painful experiments directed at remedying the suffering of humans, even if the human suffering is greater. Whether this right is or is not thought of as belonging to animals, many would think that such a right on the part of humans is adequate ground for outlawing any kind of compulsory experimentation on them (such as, for instance, by using the mentally defective as experimental subjects). There is appeal to such rights also by many who are against such things as abortion, and experimentation on embryos.

Thus both the notion of justice and the notion of rights have been thought of as pointing to moral limits on the exploitation of some for the benefit, even the greater benefit, of others, limits for which a consistent utilitarianism could have no adequate respect. I shall now try to deflate some of these objections by showing the important place our kind of utilitarianism has for the concepts of rights and of justice. I shall then go on to say something about the meaning of 'ought', and in the next chapter about moral character, in order further to develop our account of moral judgement and to meet other possible objections to it.

3. The best way of understanding the statement that someone has

a right to something seems to be to take it as the claim that there are grounds for complaint on their behalf if they do not have it, grounds for complaint against some person or persons who should have acted in ways which would have helped them have it, or have avoided actions which helped to stop them having it. Rights are of many different kinds. There are rights to try to do things, without certain sorts of interference, there are rights to be assisted to having certain things, and there are rights not to be subjected to certain forms of suffering and indignity. Moreover, since it seems that one person's rights must be another person's duties, whether not to interfere with him in certain ways, or whether positively to assist him in certain ways, the nature of the right is not made clear until it is made clear who this other person or persons are and what their relevant duties.*

Utilitarians have tended to take a dim view of rights. There are perhaps two main reasons for this. First, it is supposed that those who talk of a right see it as a reason why other people ought to act or refrain from acting in certain ways. Utilitarians have thought this a mistake. Talk of the right is at most a way of formulating claims about what these other people should or should not do, it cannot be a way of giving any genuine support to such claims. The reason why one ought or ought not to do certain things is always ultimately a matter of what will best promote happiness. This is disguised from us by the language of rights with its suggestion of some further moral reason for doing things. Second, some utilitarians have seen the conception of a right as a device by which people resist putting up with things which it is for the general happiness that they should put up with, for example compulsory purchase of their property for a new road.

The first objection has some force. On the other hand, the notion of a right can be useful when it is facts about the person said to have the rights, rather than about the person who has the

* I ignore what have been called 'liberty rights' where to speak of someone's right to do something is merely to say that it would not be wrong of him to do it, and he should not be criticised if he did. My concern is rather with what have been called 'claim rights'. However, I am not sure how sharp the distinction is, since even a 'liberty right' seems to imply some claim not be interfered with. For a convenient account of this and related distinctions – made originally by W.N.Hohfeld – see WALDRON pp. 54 et ff.

obligations, which need to be highlighted. For example, the claim that children have rights focuses attention on morally important features of children in a way which saying that their parents have obligations does not. And the claim that we all have a right to freedom in the expression of opinion focuses attention on the role of autonomous thinking in human fulfilment and on the free play of intellectual influence requisite if a person or a group's interests are to be kept sufficiently in view by society at large. As for the second, if the notion of a right is a device, it is one which can be used for good purposes, by utilitarian standards, as well as bad ones. The notion that there are certain basic human rights has surely been a force for good rather than evil, though it may be admitted that it is sometimes used as a way of pleading for privileges which it would be for the best that people should lose.

There is no reason why the utilitarian should forego the advantage of being able to express moral positions by way of talk of rights.* For if rights are understood in the way just indicated, there is nothing in utilitarianism which counts against their acknowledgement. And this is not some specially cooked up notion of rights, but one which captures, so far as I can see, its main ordinary meaning.

That utilitarianism seems to have little room for rights rests on the misunderstanding that the utilitarian is not personally concerned with the joys and sufferings of individual human beings, or animals, for their own sake, but only as elements which swell some great impersonal aggregate of welfare or distress. The very abstract level at which the basic doctrines of utilitarianism inevitably move can easily give this impression, but it is nonetheless a misunderstanding. The utilitarian proclaims the badness of suffering in all its forms, but that means that what is bad is individual people suffering each in his own particular way. Similarly, *mutatis mutandis*, with joy. It is bad that some prisoner is being tortured because of the specific awfulness of his going through that, not because it swells the aggregate of suffering in the world. If a utilitarian then battles against the ill treatment of

* Obviously he will not accept moral rights which are explicitly defined as making claims which go beyond what can be justified in utilitarian terms, as by David Lyons in WALDRON. My claim is that the language of rights is suitable for expressing and reinforcing utilitarian convictions, not that all types of right are acceptable to the utilitarian.

prisoners, it is because he sees ground for complaint on their behalf. To think otherwise is to suppose that the utilitarian is concerned only with his formula, and not with the very various realities it covers.

To say that an individual's right to something consists in there being grounds for complaint on their behalf if they do not have it, covers both the case where they themselves can make those complaints, and those in which they cannot, but others can. Some think that rights are properly ascribed only to those who can assert (or perhaps waive) them. Thus it is denied that animals, and sometimes even that babies, can have rights. Even if there is a usage of 'right' in which this is true, there seems every reason to prefer a broader sense, else we draw a false contrast between similarly wicked actions. The prime reason why it is wrong to torture a dog is just the same as the prime reason why it is wrong to torture a human being, namely that it hurts, to put it mildly. This is bound to be lost sight of if one says that a dog does not have rights because it cannot lodge a complaint on its own behalf and will make it more difficult for us to complain on the dog's behalf. (Anyway, there is an obvious sense in which the dog will complain on its own behalf, but human complaints are likely to be more effective where complaint is most needed.)

The utilitarian is not concerned with suffering as some strange abstraction which exists outside the lives of real individuals. It is the badness of certain states of individuals which perturbs him. The active utilitarian propagandist for reform is concerned with the sufferings of actual individuals and it is for their sake and on their behalf that he complains. The language of rights is a good one for focusing attention on what it is about actual individuals which is of concern, namely that they are creatures whose capacity for suffering and joy is not being taken account of by the practices being criticised.

It is true that the utilitarian is, in principle, prepared to allow that in some cases it may be necessary for some to suffer to prevent worse or more suffering for others. But most proponents of rights will also allow that this is sometimes so. What they want is not some absolute bar upon sacrificing the interests of some for the sake of others, but a strong warning sign post in the way of it. But utilitarians should wish to erect such sign posts also. If you are dealing with real suffering it is always very bad that it should be inflicted. It should always be our first aim to resolve situations of conflict by finding a fresh approach in which all benefit.

Once grant that sometimes the interests of some must be sacrificed to the interests of others, it is not clear how the notion of rights can be used to supply the fundamental answer to the question when this is, and when it is not, legitimate. The kind of utilitarian approach we have indicated seems the most intelligible one.

Nonetheless, there is every reason to institute some legal rights which put an absolute bar on sacrificing individuals in certain ways. And I believe we should go further and say also that the most desirable constructed moral world is one in which certain rights are thought of as absolute, or at the least as well nigh absolute, because over-riding them is felt to be so bad, both in itself and in its consequences, that the only sorts of cases in which it would be admitted that they might legitimately be over-ridden, because there was genuinely no alternative way of preventing some evil of quite fantastic magnitude, belong more to the realm of philosophical fantasy than the real world. And even if we look at the matter temporarily from a stand outside any such moral construction and try to think of the impersonal truth of the matter we shall surely conclude that in any individual case the desirability of keeping the strongest sentiment of respect for such rights alive in others and in ourselves is likely to be a reason for not overriding them which will outweigh proposed grounds for overriding them in particular cases.

The establishment of such legal and moral rights is possibly the only way of promoting certain great goods of personal security and keeping some of the worst horrors of life at bay. Utilitarian defences of rights along these lines are often suspected of being attempts to save the utilitarian from the unpalatable consequences of his own doctrine. But it is utilitarianism itself which tells us that certain social developments would be unpalatable and which points out the urgency of preserving sentiments which stand in their way.

So it seems that purely utilitarian considerations should persuade us of the desirability of working for or sustaining a general form of life in which it is accepted that there is a well nigh absolute bar on sacrificing individuals in certain ways. Such bars need to be supported both by legislation and by moral feelings and habits which are most naturally expressed in the language of rights. Thus a society in which everyone can rely on the treatment they receive in hospital being directed primarily at their good, and not at the advancement of medical research in general, is surely going to be a happier society than one in which one is never sure of the motivations of the doctors when they advise certain treatment. Life is better when people live in societies where they have certain definite expectations of each other, and live by certain standards, which are not always being subjected to reevaluation for the good they are doing in terms of preventing misery or promoting happiness, which does not affect the point that the ultimate justification for such a way of life lies in the way in which it promotes, and forms part of, the happiness of those who live in it, and that the desirability of individual actions turns on the good they do, including the way in which they reinforce the relevant habits and feelings.

Thus the amount of suffering from which the members of a society are saved if the established practice, particularly in all government agencies, is such that the question of using torture will just never come up, because all are trained to find the very idea of using it unspeakable, is, it seems certain, bound to outweigh any good there could ever be a serious case for saying cannot be attained without its use. Reflect, in this connection, on how once the use of torture in some circumstances is treated as acceptable, it is likely to start getting used as a routine method of enforcing policy. Moreover, a society in which there are certain habits of mutual respect between all humans, such as make torture a psychological impossibility, allows feelings of solidarity to develop which directly enrich life in ways not otherwise achievable. Therefore, utilitarians should see it as one of the most important goods which a society can enjoy that its whole way of life makes the resort to torture quite unthinkable in any serious way. (It is also worth pointing out that torture is possibly unique among great evils in that it could be abolished on the basis of widespread moral reform by individuals. It contrasts thereby with a great evil like hunger. An increase in human virtue might not be enough to eradicate hunger, if not accompanied by an appropriate degree of intelligence, but the elimination of torture would follow on a sufficient increase in human virtue. This may give a basis for thinking it more appropriate to talk about a right not to be tortured than a right not to starve. It seems particularly appropriate to talk of someone's right to something where it is bad that he has not got it, and where a reform in the personal morality of others would be enough to give it him.)

Bringing these various points together I would say that the most desirable form of life is probably one in which certain rights (such as the right not to be tortured) are treated as to all intents and purposes absolute and not subjected to continual reassessment and others (such as perhaps certain rights to free speech) as only over-rideable for unusual and quite special reasons. Where a reflective mind does feel the need to think out the basis for his actions on the basis of first principles I believe it will be found that in any individual case the desirability of keeping the strongest sentiment of respect for such rights alive in others and in ourselves is likely, in the first case always, in the second case usually, to be a reason for not overriding them which will outweigh proposed grounds for overriding them in particular cases. And, what is more, for one living that way of life it will be a directly experienced good to respect these rights and evil to over-ride them.

4. The requirements of justice cover various rather different moral desiderata, particularly in the distribution of benefits and burdens and in connection with reward and punishment, where it is thought desirable not to show the wrong kind of favour to certain individuals or classes of individuals. It is often thought that certain ways of distributing such goods and evils are to be condemned as unjust even if they serve to maximise welfare. Thus it would be unjust for the police to plant false evidence against a man whose guilt could not be proved (still more if he was thought actually innocent) because they think the public peace requires that someone be convicted for a crime. Those who attack utilitarianism claim that even if more good was done than harm by such an act, it would still be unjust, and therefore at least almost always to be avoided. Similarly, making taxation bear more heavily on those from whom it is easiest to collect it may be deemed unfair and objectionable even if the balance of general happiness is thereby favoured.

There is no doubt that people do see such things as unjust, and see this as a quality of the actions which cannot be reduced to their bad effects. And it is a part of a desirable social construction, and therefore pragmatically true, that they are indeed bad, whether this badness is actually felt by anyone as repulsive or not, whilst insofar as they actually are felt as bad they are bad in literal truth. It is not merely a useful belief but (I suspect) a fact of actual experience that people suffer peculiarly painful feelings of resentment and of abandonment if punished for what they did not do, in addition to the unpleasantness of the punishment itself.

Now it seems fairly evidently desirable that such things as the punishing of an innocent man should present themselves to our thoughts as intrinsically bad and that the suffering caused to an innocent man thus punished should have a special feel of personal outrage. Yet, granted that good and bad are ultimately a matter of pleasant or unpleasant feeling, the value of styles of thought and feeling for which things appear so must lie in the superior felt quality of life in a society in which they are pervasive. That this is so seems clear. A society in which punishment of the innocent did not repel would be in all sorts of ways a wretched one.

Agreement on the injustice of punishing the innocent is virtually universal. We are less likely to agree on what is just and unjust, or fair and unfair (which is pretty much the same), in matters of taxation, pay and so forth. Here there is no one way of constructing value in which we have all been conditioned. The only rational way of resolving these matters is by attempting to decide which world of constructed values provides the best way of experiencing the world for the community as a whole, in the immediate living of it and in all its ramifications.

My approach has something in common with rule, as against act, utilitarianism. (It is still closer to what has been called motive utilitarianism.) For I am saying that the rational grounding of our objection to what we call injustice is that the rules which discountenance it, and the motives and feelings implied in regarding it as injustice, are for the general good.

The usual objection to rule utilitarianism is that if the point of a rule is to promote the general good, then it can only be sensible to abide by it in those cases where it does promote the general good. That objection is entirely just for the kind of rule utilitarianism which professes to be an alternative to act utilitarianism, but it does not apply to the kind of rule utilitarianism to which my position is somewhat akin, which professes to be an application of, not an alternative to, the basic thesis of act utilitarianism, that what matters is good actually achieved and bad prevented. Such a rule utilitarianism insists only that, since life is best in general when we live by certain rules, one of the main things to be considered in evaluating any act is how far it sustains or otherwise our living by these rules. Still more important perhaps is the way in which it sustains or otherwise our readiness to be swayed by certain motives. This is not a subterfuge for reconciling utilitarianism and every day morality, as is sometimes alleged, but the almost inevitable conclusion of serious act utilitarian thinking.

The somewhat similar point I am making is that if we are seriously committed to the value, from a utilitarian point of view, of a way of life in which we learn to experience the world in certain moral colours, we will want to live at all times in a manner which sustains this way of experiencing the world. However, there is more to it than merely a kind of keeping in training, since for those who are once engaged in living a certain way of life, it is bad as a matter of actual feeling, not merely of construction, that things are happening which are bad in terms of their shared construction. Thus it is bad in the most ultimate and literal way for us as we are, and as it is good that we are, that the world should exhibit injustice.

The reasonable utilitarian ideal, I repeat, is not that we should be checking our every decision, made in the light of a certain system of values, by considering the ultimate utilitarian good promoted by life according to these values. If one's policy is to live in a certain way, then most of the time one should live in that way, only moving to the abstract considerations in support of that way of life when something suggests the need for further critical thought about the whole pattern of feelings in which it consists. Judged from an impersonal point of view, it is best that one should on each occasion do what promotes the best, but this does not mean that it is best always to think about things in this way. Just as there is a time for engagement in one's own more personal projects, without thinking about the impersonal best, so is there a time for living in terms of the system of values one has found satisfactory without doing so. One cannot engage in any project without giving oneself up to what it of its nature demands, rescinding from the ultimate reasons for engaging in the project. That goes for the co-operative project of living in the light of a certain system of constructed values, as much as of any other project. It should be borne in mind also that the ultimate value of living in a certain constructed value world is not merely some extrinsic good which it promotes. The very way of experiencing the world it represents may itself be a great good, in very literal truth. Thus some strange code of honour, which seems hopelessly unutilitarian, found in certain societies, may have its very own particular value, in terms of the good of actually seeing things in its terms. There is every reason to seek to sustain aspects of our own cultural heritage which may themselves be parts of a way of life, better in the living of it than any we might find by trying always to think directly in utilitarian terms.

In making the point here and elsewhere in this book that it is a utilitarian good that our sentiments are not exclusively utilitarian, my position has some similarity to points about different levels of ethical thinking made by R.M. Hare in his *Moral Thinking* (see especially chapter 3). I do, indeed, go along with a great deal of what he says there. However, apart from the fact that the basis of his thought is somewhat different from mine, I do not think that he is concerned to stress, as I am, that what in themselves are non-utilitarian ways of moral thinking bring fresh values into our 'life world' which often deserve to be cherished for their immediately felt value just as do many other cultural creations. Thus, when functioning as utilitarians, we should think of the best kind of non-utilitarian moralities, not merely as devices, nor even merely as beneficial habits of feeling, on the basis of which we will usually arrive at the correct utilitarian result which a superior being would reach more directly, but as the home of values which could not exist without them.

If the admission, or insistence, upon such points as these is thought too far from the main tradition of utilitarian thought to be called 'utilitarianism' that is, after all, only a verbal issue. I call my position utilitarian because it accepts the fundamental utilitarian insight that what really matters is how much there is of experience which is felt to be good in the actual living of it, and little of experience which is felt to be bad in the actual living of it.

5. I conclude with a few remarks upon the meaning of 'ought'.

It seems to be the case that many people experience the world as containing an *ought* which cannot be cashed in terms of the undesirability of the consequences of acting in a contrary way. Something of this sort may, indeed, sometimes be an aspect of those sentiments which are especially prompted by talk of certain rights as absolute, or nearly so, and of which I have stressed the enormous value. However, I believe that for most people these sentiments consist in a feeling of the intrinsic badness of behaving in certain ways to another human being coupled with a deep conviction that in the real world there is never reason, even in extreme situations, to believe that doing so is the only means to prevent an evil worse than they themselves immediately involve, when we take into account the sufferings of the victim, the corruption of the perpetrators, and the uncertainty of the future. Such horror at the use of certain revolting means, and conviction that their eschewal will never finally diminish the value present in the world, is both intelligible and one to be recommended on our account.

But such an account does not seem to cover all cases. For example, someone may believe that he ought not to lie, even though it is quite obvious to him that the consequences would be worse than if he did. (A sentiment against telling lies is highly desirable, on utilitarian grounds, but the occasional white lie is scarcely a corrupting evil like torture.) Can we square this fact with the account we have given of the nature of moral judgement as turning on the ascription of value qualities the idea of which goes back to the pleasurable and the unpleasurable in our experience?

There is no problem for our general account of what moral judgement is, in allowing that people often do feel actions – or more deeply the state of the will manifested in actions – to be bad, in a way which is not merely a matter of thinking their consequences bad. For one thing, the actual doing of an act may really and truly have a felt badness for an agent whose 'life world' is of a certain sort. That such actions – or the states of will manifest in them – are imagined as having such badness even when they do not feel bad to the agent is simply another illustration of the truth that value qualities are thought of as occurring even when they are not experienced, and sometimes even as occurring where what is actually experienced clashes with them. Nonetheless, the only value qualities we can encounter, imagine or make any finally coherent sense of, are qualities of the same generic kind as pleasure and pain which can only really exist as elements of sentient experience. Thus the kind of badness which we find in an action which we could only do with a sense of guilt is thought of as pertaining also to actions which are done with no felt guilt by those we think of as morally blind.

However, as we have seen, statements about what one ought to do may be more immediately imperatival or stimulatory, than other value judgements, formulating the instructions for ourselves and others which we derive from our envisagement of the moral or other value qualities of the actions and their results which we are considering. Similar remarks apply to 'should'. Not that the distinction between ethical language which directly stimulates to action and that which expresses the envisagement of values is at all sharp. Other words than 'ought' may function in the more purely stimulatory way, while all value judgement is imperatival inasmuch as the envisagement of things as good or evil is one with an inclination of the will to forward or hinder them.

Doubtless for some people judgements about what one ought to do have an imperatival sense because they express the instructions of a deity, but even then the more ultimate source of the instruction lies in the moral quality of the Deity and of actions which would go against the will of such a being. Where the 'ought' is of a peculiarly strong moral kind, it seems that the judgement that one ought not to do something is typically an instruction to oneself not to do the action, based especially on a sense of a moral quality of wrongness or badness which would pertain to the act of doing it.

This wrongness consists in the peculiar unattractiveness which we imagine in a state of mind showing insensitivity to morally significant values. We have, indeed, already contended, at the end of chapter six, that these are best understood simply as being the values to which it is peculiarly bad for one to be insensitive. There is nothing viciously circular here. The mode of feeling which discriminates certain values as peculiarly moral simply is the mode of feeling for which insensitivity to particular sorts of consideration has a specially repulsive quality which we can call *being wrong*. Thus we do not feel it as peculiarly bad that people should be insensitive to the beauty of a great cathedral, but we do feel it as peculiarly bad that someone should be insensitive to the suffering of a child in his care. In our society the tendency is to think that the one sort of value to which it is peculiarly bad to be insensitive is that which consists in the suffering of others. Of course, no one is intrinsically 'other'; he is only other to others. But it is the sufferings, and to a lesser extent the joys, of people other than himself to which it is thought peculiarly bad for someone to be insensitive.

Some think that this is a peculiarity of modern ethical feeling and one not to be entirely commended. Followers of Nietzsche may think it just as bad that people should be indifferent to great human achievements or to the development of their own potentialities, and that our sensitivity to these is as properly called moral as our concern with the basic elements of the welfare of others.

As against this I am inclined to think that modern feeling is right in ascribing a peculiar badness, worth singling out as moral badness or wrongness, to insensitivity to the welfare, especially the suffering, of others. The evil brought about by such insensitivity, and the good achieved by the kind of sensitivity to the welfare of others which stands in contrast to it, is so great that it is well they should appear to us in distinctively horrible or glowing colours. (Whether these colours are in any sense there other than for the onlooker will be discussed in the next chapter.) But I agree that if this is the sole sort of sensitivity which is encouraged important values may be lost sight of, and that we should not distinguish too sharply between this moral sensitivity and sensitivity to the value of great achievements and the development of our powers. The remarks which follow apply however precisely moral sensitivity or its lack is distinguished from other desirable traits of personality.

It may seem that, however much our account of 'ought' may do justice to the mode of thinking of those of a consequentialist or utilitarian outlook, it cannot account for the meaning of 'ought' in general. If the idea of an act which I ought not to do were simply the idea of an act that was bad in virtue of being insensitive in a particular sort of way, then it would seem always possible in principle that we might come to see this badness or this ought-not-to-be-done-ness outweighed by the badness of certain consequences of doing it. In that case no sense could be made of the position of those who think there are certain sorts of act one ought not to do, even if one will thereby prevent more acts of that same sort than one is going to do oneself, as for instance by lying so as to prevent several others lying. Yet some people, in thinking of an act as wrong, or one they should not do, think it wrong in a way which could not be outweighed by any consequences lying beyond the immediacies of the deed.

It seems to me that what often happens here is that the individual is so absorbed in contemplation of the moral quality of his own life that he cannot see the peculiarly moral values realized in it as simply among the very various values present in the world, but feels that they are so special as to outweigh all other values. He is a kind of moral solipsist who thinks that his own person – or perhaps that his own particular relation to God – is real as is nothing else. So to have this extremely strong sense of *ought* is to make oneself the centre of the universe.

It may be objected that this is not so, for he recognizes that the same *ought* applies to others. Just as he should not lie to stop you lying so you should not lie to stop him lying. There is no solipsism here, for it is recognized that others should give second place to the moral quality of what I do just as I should to the moral quality of what they do. Some of our duties are 'agent relative' in the sense that it is not equally obligatory on all agents to take certain things into account, but the existence of the different agent relative obligations can be recognized impersonally by all.

The matter calls for a deep phenomenological enquiry. Here I can only assert my view that what happens is that when experiencing an *ought* such as this, one is in the kind of solipsistic state indicated, but that when one steps back from it, one recognizes such solipsism as suitable for others too. In *Measure for Measure* Isabella illustrates such moral solipsism when she says: 'Then Isabel live chaste, and brother die.' The variations on this theme in religious thought are of considerable interest. Sometimes it is thought both that it is each person's duty to save his own soul, and that nonetheless there would be a certain goodness to letting oneself be damned for the sake of others.

So the kind of absolute *ought* in terms of which I am thought of as having duties which cannot be outweighed by consequences, other than those built into the definition of the act to which I am obligated, often seems to turn on a value thought of as belonging to my acts which cannot be outweighed by other values. Although some people do experience the world in this way, it is perhaps too far from any literal truth of things to be a wholly desirable form of feeling.

I believe that much the same account applies to other forms of moral feeling which clash with a consequentialist ethic by insisting (and not as a mere consequence of superficial misunderstanding) that our obligation is sometimes to do things, even when the consequences will be predominantly bad. Some believers in the sacredness of certain rules, who would not see the goodness of life according to these rules as something which could ever be outweighed by other factors, may not be so much concerned with the moral quality of their own souls, as with the moral quality of the life around them, and be simply unable to see the values realized in that here and now as among the values in the world, which might one day have to be balanced against more distant values. Nor is this kind of feeling necessarily to be deplored, for it is at least arguable that a sensitivity to value which is not at its strongest in relation to what confronts one most directly is not much of a sensitivity. The person who is ready to be ruthless in pursuit of distant goals is more likely to be sacrificing his victims to his enjoyment of his present sense of power than to a distant future which can only be presented in the most abstract fashion.

Once we get away from the extremes of these absolute *oughts* which seem to rest on a kind of moral or at least communal solipsism, we come to a range of alternative forms of feeling in which the 'ought' expresses the instruction we give ourselves on the basis both of our recognition of factors in a situation to which we would think it peculiarly bad to be insensitive, and from the badness of that insensitivity itself, which in a kind of potentially infinite regress, harmless in practice because it is only potential, becomes itself one of those factors. The more moralistically we approach the world, the more the second consideration, that of our own moral sensitivity, will loom large, but in all such cases the ought will represent an instruction drawn both from direct consideration of morally significant factors outside our will and from a sense of the badness of being insensitive to them.

Thus the sense that one ought to do something seems normally to include the thought that one's will would be in an intrinsically bad way if one did not. This is true both for those who think that the *ought* is solely grounded on consequences and for those who deny this. What is wrong with a will may be thought of as its indifference to, or its failure adequately to consider, consequences, or it may be thought of as more a matter of its failure to accord with a certain way of life. The non-consequentialist state of mind seems to consist either in an especial degree of concern with the goodness and badness of one's own will especially in terms of one's sensitivity to more immediate aspects of the situation one finds oneself in, or with the goodness and badness of the forms of life in which oneself and those close to one are engaged. However, those who think in this way need not be believers in the 'absolute' ought and thus be victims of moral solipsism, for they may admit that in those unusual cases where the maximisation of good will in the world requires my bad will, it would be better that my will should turn bad (though as a result of a higher order act of good willing). However, this is hardly a usual case. The need for such acts as telling lies to prevent telling lies is hardly great, and the main place of such cases in the world is as the far fetched examples of moral philosophers.

Certainly, at the level of impersonal truth what matters is the totality of what happens as a result of my doing something. To deny this is simply to seal oneself up in one limited set of values to the exclusion of all else. It is to prompt oneself to act on the basis of statements about what one ought to do which express one's response to envisaged values which one refuses to confront with other values which might have a conflicting influence on one's will. However, just as with justice and rights, so with the more basic notion of *ought*. It may well be good that we should live with a sense that there are certain aspects of what one does – not adequately described merely in terms of their total predictable consequences – attention to which is primarily constitutive of a good or bad will and it is not absurd to think that this is the value it is primarily one's own concern to actualise. After all, even from the point of view of impersonal truth, it is the case that once granted we are living in the light of a certain system of constructed values, the real values of our life will include the felt value of appropriate sensitivity to these values in our will. All the same, the concern with the moral quality of our own will is something which we probably ought not to let become too dominant. We can recognize it as a better state that our will should be drawn towards envisaged good, and away from envisaged ill, in the world at large, and in this sense see that it is not the *ought* with which we ought to be exclusively concerned.

CHAPTER IX

Moral Character

> It is evident that any one who reprobates any the least particle of pleasure, as such, from whatever source derived is *pro tanto* a partizan of the principle of asceticism. [This is a perverse principle, produced by excessive regard for the requirements of self control, which reverses the roles which the principle of utility assigns to pleasure and pain in determining right and wrong.] It is only upon that principle, and not from the principle of utility, that the most abominable pleasures which the vilest of malefactors ever reaped from his crime would be reprobated, if it stood alone. The case is, that it never does stand alone; but is necessarily followed by such a quantity of pain (or, what comes to the same thing, such a chance for a certain quantity of pain) that the pleasure is in comparison of it, as nothing; and this is the true and sole, but perfectly sufficient, reason for making it a ground of punishment. [Jeremy Bentham, *Introduction to the Principles of Morals and Legislation* Chapter II §4]

1. Bentham's view seems an inevitable but shocking deduction from utilitarianism. It implies that the pleasure some thug enjoys in bashing someone about is in itself good, and that the badness lies only in its effects on his victim. In strong contrast, such very different moral philosophers as Hume, Kant and Schopenhauer, different as their views are concerning what makes a will good or bad, all agree that an evil will (and its pleasures) is bad in itself and a good will good in itself, not merely good or bad for their effects.

While Kant's view that nothing is good in itself except a good will strikes most of us as excessive, the view that virtue is only good and vice only bad as a means, and that they are without any value or disvalue in themselves, may well seem unappealing. But are we not committed to this by the broadly utilitarian conclusions of the preceding chapters?

Our leading principle is that nothing is good or bad in itself except pleasurable or unpleasurable modes or contents of experience. On the face of it, that means that moral goodness or badness can only be intrinsically good or bad so far as they are pleasurable or unpleasurable for their possessor. One may well respond to this by saying that moral goodness and badness have an intrinsic value and disvalue which has nothing to do with the pleasantness or unpleasantness of their exemplifier's feelings and that in fact the wicked are unfortunately often much happier for their wickedness than the good are for their goodness. Although I will suggest that the assumption that for the utilitarian any intrinsic goodness or badness pertaining to moral character must consist in its pleasurable or unpleasurable felt quality for its possessor is an over simplification, it will be useful to see what conclusion it would be fair to reach on that basis.

I have suggested that a morally good man is one who is especially sensitive to morally significant values, and a morally bad man one especially insensitive to them, or only sensitive to them in a twisted and perverted way. (See chapter six §10.) These values are those to which it is so especially important that people be sensitive that respect for them should be promoted if not by law then at least by a powerfully suasive public opinion. The tendency today is to think that the values respect for which is most important in this way for each of us are those pertaining to the effects of our actions on the welfare of other humans and animals, and that concern or lack of it for these constitute the most basic sorts of moral goodness and badness. Although I shall concern myself only with the value of moral goodness and badness taken in this sense my remarks may have some application to other admirable and deplorable qualities of character, whether we call them forms of moral goodness and badness or not.

The suggestion that moral goodness is an intrinsic good because it actually feels good in the living of it is liable to be scoffed at because it is associated with somewhat simplistic hymns sometimes sung to the joys of the virtuous life. But there is surely some truth in it.

One reason for this is that the openness of moral goodness to the forms of good in the world gives it a constant source of joy. True, the good must suffer more from a sense of all the horrors of the world than the bad, but this will be qualified by a valid sense of unity with all that is making for improvement in the world, while the particular petty worries of self concern, of fear for one's own skin or reputation, with the particular figure one is cutting in the world, will be so much less that there will be more room for untroubled enjoyment of what is to come. We need not pretend that the good person is guaranteed a happy life. The good will is truly good in itself provided that it is a positive enrichment to the consciousness in which it is present. Thus when one thinks of a good person one supposes, surely rightly, that one would be oneself happier if one was more like that. For an eloquent statement of all this, the reader may turn back to the quotation from Schopenhauer in chapter four.

The felt goodness which I am suggesting pertains to moral goodness does not consist in smug feelings of self congratulation. Its felt goodness may hold in the absence of any classification which the good person makes of himself as morally good. We must not identify the felt character of a state of consciousness with what is explicitly selfconscious in it. One's consciousness inevitably owes much of its felt goodness and badness to what is not thought about at all.

2. If moral goodness is thus, at least in many cases, an intrinsically valuable form of consciousness, is moral badness usually likewise an intrinsically bad and unpleasant form of consciousness?

In considering this question it is important to distinguish between what I shall call untroubled and troubled wickedness. There is a kind of wickedness which, when it is prospering, is sheer pleasure, and there is a kind of wickedness, which is shot through with feelings of disgrace or shame. Wickedness also divides (though not sharply) into that in which the sufferings of others (crazily seen as shot through with good) are directly pursued as parts of its goal, as in direct cruelty, and wickedness, such as various forms of selfishness and callousness, which simply pursues its good with culpable ignorance of or indifference to the woe it causes. (In its worst forms, such as in the lifeless bureaucratic concerns of an Eichmann, this is perhaps even more repellent than the wickedness of the Borgias and Goerings whose lives have at least a certain richness to them such as may appeal to the Nietzschean frame of mind.) Both sorts of wickedness can be troubled or untroubled.

Troubled wickedness is certainly an intrinsic evil (though it may be a symptom sometimes of incipient good when it arises from a character in growth). So just as we can be pleased in an unselfish way that someone is healthy rather than sick, so we can be pleased in an unselfish way, for the sake of the person himself, that he is good rather than bad, if the only badness available to him was of the troubled sort.

In the case of untroubled wickedness, however, our assumption seems to bind us to the view that, when it is flourishing in its evil ways, it is intrinsically good, and only bad in terms of its effects.

All fully deliberate action, so we hold, is for the sake of something envisaged as good, or for the prevention of something envisaged as bad. To aim at something, in a conscious way, is to envisage it with some kind of attractive beckoning quality which is as much an element of the thing, as we imagine it, as are any of its more purely 'factual' qualities. Our envisagement of it as possessing this attractive quality is correct in literal and metaphysical truth if it is an experience, or content of experience, which would be a felt good in the full actualising of it. It is correct in a more pragmatic way if such envisagement fits in successfully with a construction of value which pertains to a life style which is literally good in the living of it. Similar remarks apply to the avoidance of what is seen as bad.

Thus when the wicked man pursues his nefarious purposes, he is pursuing an envisaged good (or avoiding an envisaged evil) as truly as the good man. He is, indeed, misconceiving it so far as it consists in the pain of another, for that pain is something evil in itself. He may also be misconceiving the value which pertains to his own action considered as an objective event in the world of our common construction. Still, at the core of his wickedness is his own experienced activity and insofar as that is joyful, then in literal truth it is good in itself, and if he envisages it as such in advance he knows a literal truth. For we cannot deny that, say, the destructive excitement of venting pent up bitter hatred against one's kind really is good if it is joyful in the living of it.

Perhaps we should not be frightened of this conclusion. Such understanding might even give us a better way of arguing with the wicked than is common. For as long as one holds that the excitement of venting hatred is simply bad, one refuses to enter into the reality of the situation. The result is that one's condemnation of it can hardly come home to the possible perpetrators. Whether they articulate it in these terms or not, they really know from the inside the goodness of the experience, and the pale significance which the predicates of the moralist have beside this felt good will hardly do much to counter its influence.

It would be different if the moralist could say, in effect:

> Yes, the experiences you seek or enjoy in doing these things really are, in their own way, wonderful. They really do provide a kind of blessed relief from all sorts of tensions, and allow you the joy of getting your own back on this harsh world. It would be hypocrisy to pretend these things are not good. However, have you really imagined the nature of what other people live through as a result of your activities? The fear and the pain of the victim, and the wretched feelings of those close to him or her when they learn about it, are as genuinely bad in their own true nature as your excitement is truly good. Moreover, your very excitement depends upon your not bringing home to yourself what the experiences you are producing are really like. You cannot truly imagine the awfulness of the experiences you are producing and still enjoy producing them. Besides which, you will make yourself an object of hatred to the majority of people who wish to live their lives in safety and rightly see the likes of you as a menace to their welfare, and it is not nice to feel oneself regarded as beyond the pale of normal humanity by others. If you really imagined all this, you would find that the goodness of your experiences is little beside all these evils. Thus your mistake is not in recognizing the goodness of your own excitement, but in your unawareness of the cost at which it is realized.

Doubtless such a speech would receive short shrift from most of the wicked. But perhaps it expresses a truth which would go some way to stem that wickedness if properly absorbed. Even if the villain never will absorb it, it at least helps *us* to understand two things. First, we see that wickedness is not just a matter of having tastes which differ from those of the so-called good, for it really does involve a kind of blindness to truth. Second, we see that the satisfactions even of the wicked have their positive value and that it might be better to find substitutes for them, less contrary to the general good, than merely fulminate against them.

It must be admitted, or perhaps insisted on, that there are limits to the desirability of entering into the villain's state of mind in order to do justice to the good which may be realized even there. Joys which are deeply bound up with closure of the mind to the joys, and more importantly, the sufferings of others, may have their own goodness, but if imaginative participation in them has any tendency to blunt one's own openness to the wider world of fellow feeling, it is a form of sympathetic feeling which sympathetic feeling itself will check. A good which is essentially, rather than accidentally, bound up with a direction of the will towards goals which cannot but be evil is one to which our openness should be strictly limited.

3. The conclusion of this discussion seems to be that moral goodness, at any rate in most cases, is an intrinsic good. It is not merely something it is desirable to encourage as a means to the good at which it aims, but it is a desirable state of mind in itself. In the case of wickedness, however, it seems we can only call it intrinsically undesirable so far as it is of the troubled sort. More cheerful wickedness, however awful in its results for others and however repellent for those troubled at the thought of these results, is good in itself, just as Bentham thought. The pleasures of the sadistic bully go to the credit side of the balance of good and evil in the world, however much they inevitably lead to a much greater weight on the debit side.

This conclusion is not some kind of excuse for wickedness. The wicked person acts as he does because his mind is closed to the truth of things in which his pleasures are vastly outweighed by the badness of what he brings about. Moreover, we who know this cannot imitate him in his untroubled wicked joys, and the only wickedness available to us is troubled wickedness.

We can somewhat palliate our conclusion by recalling the distinction between literal and constructed value, and note that it may be good to believe, and hence a pragmatic truth that, what is felt as good is actually bad. Yet there are risks in this, for while our construction of the ordinary world of facts and values is a necessary and useful practice, it can do more harm than good to allow it to clash in this way with the literal truth about what is good or bad in itself.

4. Still, the reader may well feel, as I do, a certain unease at leaving the matter there. He may feel that we have bypassed what is really the heart of the matter, the essential loveableness of moral goodness and hatefulness of moral depravity, in such forms as direct cruelty or stupid brutality.

So perhaps the sense that these qualities of character are intrinsically good and evil quite apart from their actual effects and even more importantly quite apart from how they feel to their possessor expresses really our belief that they are intrinsically loveable or hateful. For love, we may rightly say, is an intrinsically good experience so that the loveable in character ministers immediately to one of the highest goods of human life. The vicious, in contrast, are essentially unloveable and upon the whole the contemplation of them is unpleasant and an ill of human life.

What does it mean, however, to say that virtue and vice are loveable and hateful? Does it mean that they *can* or *tend to* be respectively loved and hated, or does it mean that they *ought* to be so? Ambiguity on such a matter is usually thought a source of great confusion.

I suggest that these expressions are best taken in a sense which straddles this contrast, as meaning that they are such that love or hate will tend to increase with increasing knowledge of their nature. If so, then even if not exactly intrinsically good and bad themselves, they are at least intimately related to what is so, inasmuch as they supply the intrinsically good feelings of love and hate with stable objects and are not mere means to some end quite extrinsic to their own nature. But perhaps one can go further and say that as objects of emotions which will grow more steady the more they are contemplated they actually are in a certain sense intrinsically good or bad just as fine or hideous paintings and buildings are.

To come to grips properly with the matter we must consider what exactly it is that is loved in love and hated in hate. Of course, it depends on the type of love. What is relevant here is the kind of love which is directed at the personality, and in which physical attractiveness is of little significance, except insofar as it is known, or believed on reasonable evidence, to be genuinely expressive of that.

I suggest that such love is normally directed primarily at what I shall call the persona. (The term is meant to suggest an actor on the stage of life, not a mask which hides the true self.) This is not the inner core of the self but an objectified version of it which cuts a figure in the common constructed world of all of us. My persona is neither my inner stream of consciousnesss, nor the volitional core of that, but something which my activity in the world constitutes as an element playing its role on the public stage of life. The persona is one of the things out there just as the houses in the street are and as organizations are which all think about and form opinions of. All who know one have their own perspectives on one's persona, and one has one's own perspective on it too. Some of these may be fairer than others, but the truth about the persona is an ideal limit on which these perspectives would converge if appropriate reflection and enquiry went on indefinitely.

Personas can have intrinsic value in much the sense that works of art can. If I love you, then my version of you is one of the true goods of the world, and if enquiry would sustain and promote that love in all who come to know you, then we can properly say that your persona is out there in the world as a generally accessible object of contemplation as something good in itself. Thus so far as moral goodness is one of the things which make most for a good persona, it is good in a way which is certainly not that of a mere means. Rather is it one of the presences in the world of human experience which make human experience worth while.

However, difficulty looms when we reflect on the way in which the most appalling (as we think) personas have been adored by masses. Do we merely discount them on the basis of a vote which we trust will go our way? The adulating masses at a speech by

Hitler could perhaps have ended up the majority.

The answer is twofold. First, the loveable Hitler of some doting Nazi's adoration is indeed a little bright spark in the world. Only if we recognize that there was something good there will we understand the Nazi phenomenon. But that little spark of good was so bound up with what was bad that it counts for very little weighed against it. Moreover, the loveable Hitler could not occur except in minds subject to delusions appalling for the human race and eventually for most of them. It is not just a good which does immense harm but one which could not exist except in a state of illusion, which is as much as to say that the Hitler persona was not truly good, with that truth which pertains to a construction which stands up to enquiry. A really loveable persona is one which will become more loveable the further enquiry and reflection go, for it will be bright with the joy which stems from the very nature of the will of which it is the public symbol. In contrast stand personas which will necessarily become more hateful the more fully they are presented, being the symbols of wills whose very nature it is to produce evil.

5. But cannot love be directed at the core of the inner self existing in a person's individual consciousness which is the final source of his actions and of the persona he thereby creates as an object in the public world? And does not the question still remain whether that inner self is not intrinsically a good thing when it is virtuous and bad when it is vicious, in a manner which cannot be equated with its being pleasurable or unpleasurable in the way it feels?

It would be a mistake to distinguish sharply between love for the persona and love for the inner self. For when the persona is really intimately known it has an intrinsic fit with the inner self. They are not quite the same thing, but the relation between them is so close that delight in one amounts to love for the other. The beauty of a work of art presented as a component of consciousness supplies no kind of intimation of, and in no way has an intrinsic fit to, the complex of atoms which is working on us to produce the experience of it. In contrast, the moral beauty or ugliness of a persona, while it is not exactly a depiction of the quality of the feelings and volitions which actually occur in the consciousness which is constituting that persona as object in the world for us, certainly has some kind of intrinsic fit with it. Moreover, our feeling for the persona is somehow directed through it at the inner self in a way in which our feeling for the work of art is not directed at the system of atoms which, at least on a conventional view, is the reality immediately prompting my experience of that work.

That, I believe, is true. But it still leaves us with a problem. The persona as object for others certainly has properties which are not literally present as qualities of the inner self. And it seems that the kind of essential loveability of a good persona is not normally there in the inner self. Indeed, it could only be there as a form of self complacency which may perhaps be contrary to the very sort of inner self to which the good persona points. The matter is difficult, however. The more one sees people live out their lives on television the more one is puzzled as to how far qualities of character cease to be genuine when they are self consciously enjoyed.

The problem I have in mind may become more obvious if you consider the quality of cuteness. A child may say things which give it a cute persona. This cuteness is a property which the persona, as object revealed to those who look at it, really possesses and it may do some kind of justice to the inner self of the child. But if that inner self actually has felt cuteness as an element of its own nature, then the persona was misleading. In short there are personas which do justice to the inner self only so far as the quality of the persona does not actually pertain to it. Now may this not be true of the loveability of certain sorts of good people? Their loveable persona would give a false impression of their inner self if the loveability were there as a felt quality. But that suggests that love can only be directed at the inner self by misconceiving it as possessed of an intrinsic goodness which cannot really be there. For intrinsic value only pertains to what feels good, and in this case it is essential to the moral goodness that it should not feel good in its actual exercise.

Perhaps the difficulty will seem less if we enquire a little more deeply into the nature of that inner self which is the ultimate home of moral goodness and badness, as indeed of other qualities which contribute to the true value of a person. In doing so I must anticipate ideas to be stated more fully in the next chapter. For the present it will be sufficient to explain that I conceive the self as an essence which is as it were seeking to actualise itself as fully as possible in a certain stream of consciousness and via that upon the world beyond. Otherwise put it is a kind of teleological law governing a stream of consciousness so that its activity will do what is required to actualise a certain complex form in as full a way as circumstances permit.

We have seen that consciousness as-it-were strives to free itself from distress and tension and to make its states as pleasurable as possible. In us this largely takes the form of striving for experience by which we sense or represent the things and persons around us in pleasant rather than unpleasant colours, thereby constituting certain external situations as our goals. But not every consciousness is concerned with the same forms of good and evil. Each is tuned to bringing forth, by its own activity, situations which will feel, or be conceived as, good rather than bad, in a manner which, taken all together, is peculiar to it. Thus the inner self seems to be a certain system of values which strives to actualise itself, in the first place within that particular consciousness and, as a result of activity originating there in the world beyond.

In a certain sense this essence which is our inner self is bound to be a form of the good inasmuch as its actualisation will take the form of various felt or represented forms of good. Its full actualisation, unless it contains essential contradictions within it, will take, so far as the self's own consciousness goes, the form of pleasurable feeling while its activity is that of a particular system of values trying to actualise itself. That actualisation is in the first place an actualisation of those values within its own consciousness but it spills over, so far as it is effective, into their realization in what goes on beyond it.

If we love someone, in the deepest way, we cherish the form of the good which we sense as being the inner essence active in them. We sorrow especially with the sorrow of the beloved because it is that particular form of the good which is failing to find its full actuality and exists only in a defective and sorrowful form.

Such love can encounter its object in a way which goes beyond mere representation insofar as the form of the good, which is the other person's inner core, can be actualised imaginatively in one's own thoughts. If we do not merely imagine it, but come rather to instantiate it ourselves, love has moved towards some kind of identification.

We can now see why virtue is, after all, something intrinsically good. The virtuous person's inner essence is a form of the good in which enjoyment of the happiness of other people, and satisfaction at what relief of their sorrows can be brought about, has a central place. If the form of the good constituting the virtuous person is realized with some fullness, it is so in the form of joy within himself at the joy of others. However, we draw an inappropriately hard and fast line between the self and the world in which it is operative if we do not regard the very joys of other persons which its activity produces as that self actualised within the wider world. Thus we can say that the full realization of the morally good self will or would consist in intrinsically good states of affairs both within its own consciousness and beyond. Moreover, it includes the actualisation of a certain persona as an intrinsically loveable object accessible to the enjoyment of many. Of course, the good self is normally condemned to live in a world of suffering and its consciousness will be full of sympathetic griefs. However, the actualisation of that self in the world beyond will be in the form of an improvement of it and there will be a sympathetic feeling of pleasure at that improvement within its consciousness, as at something with which it will rightly feel itself especially associated as it is not with the evil in the world.

Actually we might well distinguish two aspects which constitute a person virtuous. His inner essence must be a form of the good in which the welfare of others plays a large part, but it must also be an essence which is not too frail in imposing itself on situations. Moral badness may take the form of such frailty rather than of anything essentially bad which would pertain to actualisation of the self. Not that there is any sharp distinction here, since frailty in exerting an influence is likely to stem from some defect in the form of the good in question.

Thus a virtuous character is a form of the good as it were attempting to impose itself on the flux of consciousness and on the world around it. The intrinsic goodness of the virtuous self is the intrinsic goodness or pleasurableness which would pertain to situations which realize it with any fullness, situations including the figuring of the persona which represents it on the public stage as an intrinsically delightful object of contemplation. There is a sense, then, in which the good self is intrinsically good.

How do these reflections bear on the morally bad self? This like the morally good self is a system of values trying to impose itself on the world. There must be something positive about these values else they would not reinforce or attract. On the other hand the good in question is intrinsically bound up with evil beside which its own element of good pales to insignificance. If this evil is registered within its own consciousness, as in troubled wickedness, it will be a system of values in direct contradiction with itself and such that it is truly bad at its very source. If the wickedness is of the untroubled sort there will be a certain kind of goodness at the heart of the self's activity. Yet to suggest that this means that the self is intrinsically good is to make an inappropriate distinction between the actualisation of the self which lies within its own consciousness and that which lies beyond. The self of Hitler was actualised more in the Germany he created than in whatever personal joy he took in it all. That self could not be actualised with any fullness except in an intrinsically dreadful reality. This includes, the hatefulness of the Nazis as personas on the public scene for those who are not caught up in the delusions of that moment. Such intrinsic goodness as pertains to the consciousness of the triumphant ill doer, moreover, can only occur there in an untroubled form because his eyes are closed to the truth of things. Did he realize what he was, an essence the actualisation of which takes the form of an intrinsical worsening of the world, he would hate himself. Thus a wicked self is one whose actualisation in the world is necessarily in the form of what it itself must join with others in hating so far as it understands itself and in that sense it is itself something intrinsically bad and what no one could knowingly wish to be.*

I do not wish in talking of the self thus to suggest that character cannot change. However, character change, without loss of personal identity, is surely the persistence of some core aspect of

* For an important philosophical discussion of wickedness see *Wickedness: A Philosophical Essay* by Mary Midgley, London, 1984. The author sees some wisdom in the old idea of wickedness or evil being something essentially negative or a lack. The ideas of the self suggested in this chapter owe much to such thinkers as Spinoza, Santayana, and Bradley and to Charles Taylor's views on teleology.

such a personal essence, while the larger self which has grown up around it, and which was only one possible specification of that core, gives way to another. At the core of a wicked or good self may be an essence which was realizable in fuller versions good and bad.

CHAPTER X

Ethics and Metaphysics

§1

The idea that ethics can, or should, have a metaphysical foundation is still contentious. Perhaps G.E. Moore's critique of 'metaphysical ethics' continues to exert an influence. Also metaphysics itself (understood as a distinctively philosophical atttempt to understand the ultimate nature of reality) is still viewed with suspicion by some. So far as there is a dominant movement of metaphysical construction today it is a materialist or physicalist one, and there has not been much attempt to relate it to ethics. However, some philosophers, for example, Thomas Nagel, and Derek Parfit, have argued for ethical positions on the basis of views of the human self.

I am speaking of philosophy in the English speaking world. Positions which are, in effect, metaphysical have been associated for much of this century (though less so now than once) with moral philosophy in France and Germany. This is obviously true of existentialism and there is some kind of ethics lurking in the background in the phenomenological ontology of Heidegger. Another story, which cannot be considered here, is that of Marxist approaches to ethics.

Yet there is something very odd in the dissociation of ethics from metaphysics. The difficulty in answering such questions as 'How is it best that we, or how ought we, to act in the world?', lies as much in deciding what *we* really are, what *action* really is, and what the *world* really is as in deciding what *ought* and *best* are. Even these last problems have a metaphysical dimension, but the first simply are metaphysical problems.

As a sample of fundamental questions about or in ethics which can only be answered in the light of a metaphysics, take the following (not all of which we can discuss) where the true nature of the referent of the italicised expressions is a large part of the problem: How far are *human foetuses* and *embryos* sufficiently the same sort of things as *full human persons* to be thought of as having similar rights? How far are *animals*? Can *natural objects* other than *humans* and *animals* have any value not a matter of their effects on humans and animals? Is humanity a *totality* which can be regarded as having interests? Is the *equal pleasure* of *one conscious individual* always an adequate substitute for that of another?

Metaphysics has peeped through at times in the earlier chapters. But in this final chapter I shall give a brief indication of some positions in metaphysics which I believe can be established independently of ethical considerations but which can also help consolidate and develop the views which have been advanced about the foundations of ethics.

§2

There is a certain totality which constitutes the whole of what it is like to be you or me at any particular moment of our conscious existence. It can be called by such names as our present state of consciousness, or present whole of experience. Whatever we call it, it is the whole of what would be before the mind of someone who could imagine just what it was like to be us then. The most prominent part of it is our own felt bodily being and behaviour in the world as it is 'lived' by us, and the immediately presented perceptual and emotionally drenched world in which it takes place, together with a flow of verbal and other symbols, and of emotions, which are not attended to as objects in their own right because they constitute our thoughts and feelings about what is not actually present in experience. There is also much by way of thought, sensation, and feeling which is only there in the dim background of consciousness, but which forms a considerable part of what it is like to be us at that moment.

Our experienced or felt lives consist in a series of such total experiences which flow into one another – what William James called our stream of consciousness. Some philosophers have supposed that all that I myself can strictly *be* is my stream of consciousness, while others have suggested that each total experience is a distinct I, and that there is no persisting identical self.

These suggestions are closer to the facts than either of the two main alternatives, that I am my body, or that I am a non-physical entity not given in experience but which somehow has the experiences. The first view is objectionable, because if my body behaved and made noises just as it does, and all that goes on inside it of a physically detectable sort was just as it is, but, *per impossibile*, there were no associated stream of consciousness, then there would surely be no real I. There would certainly be nothing deserving the kind of concern proper to a person. The second view is objectionable, because, provided the stream of consciousness is as it is, it cannot matter one iota whether any 'pure ego' is in the offing or not. True, both the materialist and the believer in the pure ego may hold that there could not be a stream of consciousness without the object they regard as the true person. But even if this were so, their relevance to the existence of a person would only be as necesssary conditions of the kind of stream of consciousness necessary to sustain one, for to think of a person's existence as still an open question, once granted that there is a stream of consciousness fitted to be his, seems absurd.

This is not to say that the stream of consciousness is the self nor to conclude that there is no one and the same enduring self but only a constantly changing process. For there to be a self there must be a *distinctive way of consciously cognizing and dealing with the world* present at each moment in a stream of consciousness (or at least in the offing, in the sense that the total flow of consciousness is continually emerging from and returning to it) and there seems every reason to say that this is precisely what the enduring self *is*.

After all, in what sense is anything which exists over time, the same thing throughout its existence? What, in general, is a continuant, that is a particular thing, in a quite ordinary sense, which exists over a period of time and which unlike a mere process is wholly there at each moment of its existence? My view is that my house (say) is the same house at different times because it is a form or universal which is actualised, in each of a series of momentary phases which take the place of each other, and occur in a particular part of the world specifiable in terms of the objects with which they are most closely interacting, these objects also being universals actualised in a series of momentary instances.

Surely that which is the same in different parts of the universe must be a form or universal. The same wallpaper, as a particular, cannot be on two walls, but the same universal pattern can be. But different positions in time seem to be as much different parts of the universe as are different spaces (or different experiential wholes). If that is right (a matter I cannot argue further here) it follows that continuants are universals of some kind.

A continuant is, however, to be distinguished from a mere abstract universal, like a mere quality or pattern, for it is a universal of the type which has been called concrete. That is, it is a universal, which would otherwise be merely abstract, considered as tied down to a particular set or series of causally related instances, where the fact it is present in one instance is intimately due to the fact that it is present in others of the same set or series. The same abstract universal form (the precise structure and the precise kinds of material used) could, at least in principle, be present in two different houses, but the houses would be different because the abstract universal would be realized in two different concrete universals, each tied to a different set of causally linked instances.

Of course, houses change a bit over time. So by the continuing house we must mean the most specific universal that is realized *all along the series*. (We will not, of course, normally have full knowledge of just what that most specific universal, realized all along the line, is, but the grounds for thinking that there is a certain continuing individual are grounds for thinking that there is some such universal.) If that is the only series in which that universal is exemplified, we can call it more *fully* an individual, than if there are plenty of other such series.

Thus to interpret the self as a distinctive form of consciousness, present all along the line (or usually there and always in the offing) is not (as sometimes supposed) to regard it as a continuing individual only in some second rate way, but as so in the only sense in which one and the same thing of any kind whatever could conceivably exist at different times.* The self like a house, is a concrete universal. True, there may be times when it is convenient to say that it exists, even though it is really only *in the offing*, but (quite apart from the fact that this may be true of physical things also) that does not dent the reality of the identity when the essence *is* actualised.

A more important comparison concerns the fullness of the individuality of different continuants. Upon the whole it would seem that human selves are more fully individual, much less susceptible of duplication, than physical things such as houses. If there are types of conscious creature whose styles of being conscious duplicate each other they are less fully individual than humans seem to be. Even so, the self of one would be a different self from that of another, for this same style of consciousness would be differentiated into distinct streams.

Some have speculated on conditions in which streams of consciousness might merge or bifurcate and personalities get mixed up. I do not rule out such possibilities, nor deny their ethical interest, but they only show that identity across time can be indeterminate, not that personal identity is not normally as firm a fact as fact can mostly be. As things are, each of us exists as an identical self over time which is sharply distinct from others.

For it seems that each of us is a highly individual way of experiencing and dealing with the world which gets established as a definite and distinctive thing at some early stage in our stream of consciousness. Otherwise put, each of us is an individual personal essence or essence of consciousness. Moreover, personal essences such as ourselves are not just passively exemplified in a series of instances. Rather, the activity which goes on in each moment of consciousness, and which determines the effects of

* Some philosophers think that if I am a genuine entity which can be one and the same over time, that entity must either be something physical, like the brain, or a spiritual substance of some usually quite unexplained sort. See, for example, *The View from Nowhere* by Thomas Nagel, p. 45. However, the universal present in each phase of a stream of experience is no less and no more one and the same than the universal present in the stream of being which constitutes, say, a neuron across time. So the need to find a substance, in the only intelligible sense, need not drive us to identify ourselves either with our brain or an unexperienced soul substance.

that consciousness on the world around it, and on the further phases of its own stream, is, to a great extent, a doing of what is required in order that that personal essence shall continue to be reactualised, and reactualised in as full and dominant a form as possible. In short, my activity is in large part an effort to cope with the situations which confront me in such a way that my personality does not go under, but is, rather, realized in an increasingly complete form both in my own experience and in the world beyond. This is not just a matter of my being able to keep up my particular styles of thinking and behaving. It is also a matter of the realization in the surrounding world of the kind of thing it is distinctive of my personal form of consciousness to find good, and the diminution of that which is distinctively bad for it.

As a result of the unfortunate way in which consciousness is too often spoken of by philosophers and psychologists, expressions like 'the stream of consciousness' are liable to suggest something rather ghostly and ethereal. This is a great mistake, for – as phenomenologists have always insisted – one's consciousness, or whole of experience, includes the most grossly physical aspects of one's being, and of the world around one's body, as it and they are for oneself, and is far from some stream of mere thoughts. To understand this better, it should be realized that a whole of experience normally divides into a self aspect and not-self aspect. (Compare *Appearance and Reality* by F.H. Bradley pp. 75–82, and my examination of his view in MANSER AND STOCK.) The self aspect is the activity by which my personal essence struggles to preserve and enhance itself while the not-self aspect includes all that is merely there, and not felt as my own doing. The self side may consist merely in thinking, saying or writing things or it may consist in strenuous physical doings, in playing tennis, in washing a car, and buying someone a drink, in fighting, or in making love. In these cases both the activity and that on which it is directed are at once contents of consciousness and real portions of the physical world of daily life. Or this is so at least in so far as they are somatically or perceptually present, rather than so only by way of representation. You may say that the physical world is never within consciousness except by representation, to which I reply that the physical world of daily life is a construction of which the only parts with literal existence are those present as actual constituents of someone's consciousness.

There is, of course, a larger setting which makes possible the existence of streams of consciousness in which a self aspect continually re-exemplifies the same personal essence, and acts so as to preserve and enhance its being. A clot in the bloodstream or heart condition, not presented or represented within the stream of consciousnesss, may radically alter or terminate it at any moment while genetic endowment and other physical factors may largely have settled the nature of the personal essence active there. But however precarious the existence of the stream of consciousness and the self actualised within it, that self is a genuine force in the world, and its most basic concern cannot but be with its own realization, that is, with its survival in as full and dominating a fashion as possible in a world which presents it with a not-self conforming maximally with its own favoured forms of the good.

Since each of us is such a personal essence, a philosophy of life which offers us no guide in our quest for personal self realization will never satisfy, and cannot be complete as an address to genuine persons. Thus an adequate metaphysics of the self shows that a utilitarianism which bids us live always at the level of impersonal ethical truth is beside the human point. Nor is this mere human weakness. Even from the point of view of the impersonal utilitarian truth, the main values which can be actualised in streams of consciousness are forms of such self realization as they are felt there.

§3

Metaphysicians who emphasise the significance of the individual stream of consciousness are sometimes charged with being half way into belief that their consciousness is the only one or even the sole reality. How do I know that there are any streams of consciousness besides my own, or even any reality beyond my own consciousness and its contents? One retort would be that if there is a problem here, it is not created, but revealed, by such metaphysicians. Still, it will help illuminate the metaphysical basis of ethics if we consider briefly why such sceptical questions are so shallow.

In one sense I quite certainly have cause to believe in streams of human and animal consciousness besides my own for life would be insupportable if I did not believe that when I talk to, or embrace, another, there is a consciousness in or for which I figure as a person with whom another self is actively engaged. But that is only to say that it is good for me to believe this, it does not establish that I am right in doing so. However, much of what I experience comes as self evidently the expression of another's thoughts and feelings. Anything can be doubted in words, but I cannot really initially regard the words or gestures or my physical contact with another, as they figure in my consciousness, as mere passive contents of my own consciousness, or as the presentation of physical events devoid of feeling and meaning, and then wonder whether they come from another personal source. For they are experienced as the expressions of another's thoughts and feelings, and as an indication of their living in a world largely the same as mine, and it is only a false metaphysics (according to which there cannot be real intrinsic connections between different phenomena) which refuses to take them, at face value, as my side of an intrinsic connection between the way in which things unfold in different streams of consciousness. Although my experience does come as a distinct whole, it contains within itself an irresistible sense of being a fragment of a larger system of reality. The sense of communication with others is simply a case where the nature of some elements of that larger system is particularly perspicuous. Indeed, it is not that different in kind from the way in which each moment of total experience contains a sense of intrinsic connectedness with earlier phases in the stream. Each phase of experience feels itself both as the continuation of a personal series of experiences, and as an experience which is but one contribution to a system of interacting individual consciousnesses.

Once we begin to understand how the characters of different things interpenetrate in the world, and do not have distinct natures independent of how they stand one to another, we can resolve some of the problems we have seen arise in connection with the contrast between intrinsic and instrumental values. Moore only attached intrinsic value to what would be good in isolation. But no reality can ever be conceived of as thus existing. Thus its value as it is in itself cannot be distinguished from the value it owes to its relations to other things. But that does not mean that everything is good only as a means to other things in a futile regress. Rather, the value a thing has in itself is one with the value it owes to its relations. The positive and negative values realized moment by moment in the consciousness of each individual are essentially those of someone playing a particular role in a community. They could not occur out of that community nor be realized other than in the consciousness of individuals.

The fact that he is just one among many conscious beings will be acknowledged by every reader of this book who accepts that it had an author, so these doubts need detain us no longer. But although each of us inevitably has a general sense of the existence of others, we are liable to be so preoccupied with the world, including the lives of other conscious beings, as presented in our own personal vision of it, that we do not bring home to ourselves the way in which there is difference as well as sameness in the versions of the common world which figure in different streams of consciousness.

If we really do so, we realize that the good which occurs in another consciousness is as genuinely good as the good which occurs in ours, and that the bad is as genuinely bad. So long as we identify the world presented in our own consciousness with *the* world, we fail to realize that our values are just one limited set of the world's values. We cannot adequately take in the real character of what is going on in another consciousness without seeing the aspirations and fulfilments present there as having some validity of their own. For the good, in its every species, attracts and the evil, in its every species, repels whenever it is adequately envisaged. It is only lack of imagination which stops one entering into the forms of good and evil which occur in the consciousness of others, some of which may be very different from those which occur in ours.

Some form of sexual aberration, for example, may present itself to our imaginations as inherently nasty, but if we really imagined it as it is in the actual enactment of it, we would see it as the essentially good thing it feels to enthusiastic participants. Both my thoughts and a participant's thoughts, when he is merely thinking about it, ascribe a value to it from the outside. The only value truly realized in these thoughts is the value involved in having them. Insofar as my thinking of the aberration lowers the quality of my consciousness, that is a disvalue comparable to the positive value of the heightening of the consciousness of the participant when merely thinking of the activity (when arousal is not virtually a form of participation). But when he is actually involved in the activity a value is realized at the point to which both our thoughts are directed, and that is the value which is most properly the true value of the activity.

The same goes for all forms of life. They may seem unattractive to an outsider, and attractive to an insider. To realize that all streams of consciousness are equally real is to realize that values which might be realized in actually satisfying feeling all deserve an initial hearing by rational ethical thinking, since all must have their appeal for a mind which grasps their nature. However, a special importance attaches to the value found in the activity by participants at the actual time of engagement in it, for here the values realized are not simply the reflections of values, as they are conceived, in the act of conceiving them (the goodness or badness intrinsic to the thinking of the supposedly good or bad) but values actually there in the reality the conceptions concern.

There are doubtless difficulties in weighing conceived values against actually felt values, and in the contrast between the two. The real value of what we are doing may turn to a great extent on our conception of what we are up to. But the essential point is that any value actually felt within a consciousness (and not merely attributed to it from outside) is as real and genuine a value as can be. There cannot be mistake here, because here is actual exemplification of that which we attribute whenever we think about the value of anything. It is only because it is so hard to take the consciousness of others as reality on a par with our own consciousness that we fail to realize that, from the point of view of impersonal truth, every felt value is a real value. Even when we cannot actually imagine the forms of good and evil which occur as sensible presences in another's consciousness, we can realize the abstract truth that they are values as genuine as ours, and presume that what someone evidently seeks to sustain or to remove is realized within his consciousness as a form of good or evil which would be found such by anyone who adequately imagined it.

Several things commonly stop our seeing this. First, there is the fact that as persons seeking our own self realization we cannot but be attuned to our own forms of good and evil as we are not to others. Moreover, it is built into the nature of a stream of consciousness that the self is directed on what is, has been, or may be brought within its own experience, as it cannot be upon what must stay beyond it. Nothing seems quite as real to me as that which belongs to my present whole of experience, but what I remember or expect within my own stream of consciousness at other times seems very nearly as real as this. True, my experience intrinsically points beyond itself to the experiences of others, but it never brings their contents before me with the same intimacy as it can what has been present in my own past, while it is part of what it is to be an individual self that I have an especial concern with my future as it will be actualised within my own stream of consciousness.

But this natural sense of the special reality which pertains to myself, and of what I can directly confront within my own experience, may be strengthened or weakened by our more or less explicit metaphysics. For this may either reinforce or challenge the common or garden level of thought at which there is no distinction between *our* world and *the* world (meaning by our world here the things around us as they ordinarily confront us and all that more of basically the same sort from which we are only cut off by our position in space and time, or perhaps lack of some perceptual aid), and thus encourage or discourage us from recognizing the existence of values not belonging there. The advantage of adequate attention to the status of consciousness, and to the way in which what it directly confronts is its private personal world, is that we see that our version of the world is but one of its versions, and its values only a sample of the variety of values in reality at large.

Strong realist ontologies, for which we directly confront a world existing independently of us, in much the character it presents to our observation, deprive us of the opportunity of grasping how the world comes in different versions. However such straight realist metaphysics officially conceives values, it is likely to discourage us from realizing that every imaginable world is the world of some conscious subject, that there are as many such worlds, each with its own values, as there are subjects, and that the impersonal truth about value incorporates them all. The tendency of all realisms is to reduce other individuals merely to that of objects existing in our world. This deprives us of a sense both of their subjecthood and of ours, theirs because it is demoted to that of object in the world, ours because it is promoted to being the world in which all things are.

The less metaphysically minded may move quite deftly from a realist perspective to one which registers the variety of versions in which the world presents itself. But once the realist side of the two pronged nature of ordinary thought takes control, the ethically significant realization of the variety of versions in which the world comes, and of how what is for me the world is only one of them, is jeopardised.

I am not denying that different versions of the world do different degrees of justice to the one real world which includes us all with all our versions of it. Nor do I deny that to a great extent we all share a common version of the world. Sameness is never an all or nothing affair, it is always sameness with a difference.* Realism threatens our sense of the difference, but the wrong kind of idealism could threaten our sense of the sameness. Individuals who have anything to do with one another must be directed on what are substantially the same objects, and with humans, and perhaps sufficiently similar animals, the identity goes considerably beyond the minimal structural isomorphism which is absolutely required. The more we share the same cultural heritage, the stronger the identity and the more even the subtlest intellectual and aesthetic significance of what we confront will be out there in a universal not-self. Such sharing is itself a felt value for each, yet it should not stop us grasping something of the radically different values which may exist in alien versions of the world. For both our own and these alien values equally matter, in literal truth, to whatever extent they are themselves values of actual experience or are such that felt value pertains to the belief in them.

Still, the being steeped in one's own world and thinking of it as the whole of reality seems an essential aspect of what it is to be a distinct person. Personal identity necessarily brings with it a partial myopia. Since much of the richness of the world comes from our separate identities this myopia is in part a good. We do

* I am speaking of the real sameness which holds in the world, not of the sameness which is merely the product of the fact that we have two expressions referring to one thing.

justice to this good in stressing the importance of self realization. But we would be myopic to excess if we did not feel drawn to moving from time to time to the level of impersonal truth as represented by a broadly utilitarian ethic.

§4

At this level we become aware of the truth not only that anything actually experienced as possessing value, has, qua element of consciousness, that value in the only possible literal sense in which there is value at all, but that a value not felt as such within a consciousness is a metaphysical absurdity. It is, indeed, inevitable and desirable that we construct a world together in which there are supposed to be unfelt values, but the real value of such construction can only be judged at a level where its actual incoherence is acknowledged.

This utilitarian insight (so it seems to me) is just one phase in the development of the metaphysical insight that in the end nothing can be conceived as truly existing which does not do so as a form of, or element in, consciousness. Shapes and colours, movement and weight, distance and size, no form of these can be imagined which could occur other than as an element in some conscious being's experience. One cannot imagine a world except as presented to some particular type of consciousness (unless perhaps it be a world composed of forms of consciousness). Some acknowledge this, but say that all the same there *is* a single real physical world, however much we can only imagine it in terms of our own subjective forms of perception and conception. But if we can provide ourselves with no kind of picture of this physical world, what we say about it amounts only to a complex system of verbal stimuli by which we prompt ourselves to effective action in a world the actual nature of which remains quite hidden (as does also the true nature of our own action), while if we try to conceive it in any more full blooded way we can only dress it out in characteristics which are quite evidently only there for us. If one realizes this, and yet remains convinced that there is some kind of reality to the physical world beyond that of being an object of human and kindred forms of consciousness, one must interpret our descriptions of this independently existing physical world as capable of literal truth only insofar as they specify solely its structure, thus as telling us something very abstract about it, comparable to what someone born deaf may know about musical scales, or the structure of a fugue.

The idea that we only know the *structure* of the independent physical world has been advocated by many perceptive thinkers, among them Bertrand Russell, and there is much in its favour. It only becomes absurd if it is held that the world only has structure, and does not have some concrete way of realizing that abstract structure. The view that we know only the structure of the physical world implies that there are unknown qualities and relations there which conform to a certain abstract schema, and which we can pick out individually only either as those which play a certain part within that schema or as the sources of certain familiar experiences. One cannot hold this view, and then deny that these concrete qualities and relations are anything definite beyond this.

In my opinion, science genuinely does tell us something of the abstract structure of an independent world which underlies the existence of human streams of consciousness. Indeed, even the physical world as we ordinarily imagine it, and which only exists, other than as an ideal construction, as the not self of our perceptual experiences, conveys something of the structure of the reality which underlies it. So one can form a genuine conception of a world which is not merely the world as presented to human consciousness, if one sticks rigidly to a purely abstract structural account. When I buy four apples of unequal size I am putting myself into a special relation with something which can exist beyond human consciousness, comprised of four similars having degrees of something which is isomorphic with spatial size (such as the number of less complex things which somehow relate to each other so as to constitute the being of each similar). But the apples, with their pleasingly rounded shape and inviting rosy colours, with their tastiness somehow contained within them, can only exist as objects for subjects like ourselves. They exist in a full fashion insofar as they are actually seen, felt and tasted, and only as an element in a useful construction so far as this is not so.

But can we know nothing of what fills out the abstract structure in literal truth? I believe we can. In the end nothing is *imaginable* in a fully concrete way except modes, and contents, of experience. What if the reason for this were that nothing can *exist* in a fully concrete way except modes, and contents, of experience? Ultimately every claim that something is impossible rests upon an inability to imagine it. Perhaps our inability to imagine what lies beyond experience constitutes an insight rather than any kind of defect.

I do not dispute the intelligibility of things we can specify in terms of what we can imagine, but which outrun our actual power of imagination. As has often been pointed out, we cannot distinctly imagine a chiliagon. Still, we can imagine four sided figures, and we can imagine the relations of 'having one more side than'. So we can specify a chiliagon in terms of what we can imagine. As such we can be said to imagine it indirectly. Our inability to imagine the totally unexperienced is different, for we cannot even imagine it in this indirect fashion.

I believe, in fact, that our very notion of what it is for something to be genuinely actual is that it exists with the same kind of positive thereness as our own experience. Further argument for this view is impossible here, so let me turn rather to its implications. Of these the most significant is that the actuality which lies behind our four apples as we imagine them is somehow composed of experience. The existence of an apple (as 'a thing in itself') over time must consist in the occurrence of a stream of process having some very general affinity to the stream of our personal consciousness, in which a common essence is continually repeated with variations due partly to the impingement of other such streams of process upon it. Or at least, and more probably, this must be true of physical realities of some more ultimate sort (say at the sub-atomic level) whose complex interweavings constitute the being of the apple.

The picture of the world to which these considerations point is of a vast system of streams of experience. These develop in interaction with each other, each of them under the control – as it were – of an essence which they actualise as best they can under the influence of others. At the non-human level it is likely that these streams merge and separate in a way which only happens to a very limited extent at the human level. Still, at any one time there must be some articulation of that totality of experience which is the world into distinct wholes, for it would seem that an individual experience must either belong to or be an experiential whole distinct from others in something like the way in which each person's total experience is at any moment.

§5

However, this apartness of one whole experience, or one stream of experience, from another poses its problems. How can they be in any way related to or impinge on each other? Whenever we imagine things in relations of any kind to each other what we really do is think of them as each making a distinctive contribution to a whole they form together. To imagine how the streets of a town relate to each other is to imagine them as distinct elements in the pattern of a town as a whole; to imagine a system of personal relations is to imagine the persons as elements in a community with some kind of overall life.

The standard way in which we imagine all things as in relation with each other is to think of them as occupying different positions in space and time, more ultimately in space-time. But the space or time of ordinary thought is as 'subjective' a reality as are values and colours, while the space and time of science – insofar as it moves away from this subjective reality – is a highly complex abstract structure of unimaginable items in unimaginable relations. If we try to imagine the streams of experience which make up the world as genuinely related to each other, as really in a common world within which they can interact, we find we have to postulate a whole of an experiential sort in which these experiences can figure. This (I think) can only be a single cosmic or divine experience in which all the variety of the world is present in one unitary state of complex feeling. All ordinary finite streams of experience must combine to make it up somewhat as a multiplicity of thoughts and feelings go to make up a single state of finite consciousness.* Moreover, it must be a unity not merely of all the streams of experience occurring at one time, but an eternal changeless unity of all of them experienced as unfolding

* Though this absolute consciousness is inconceivable by us in any concrete way the abstract description of it can be cashed in imaginable terms. It is a variegated unity containing distinct elements in the sense in which a single state of mind such as I am familiar with in my own case is. For further explanation and argument see my *The Vindication of Absolute Idealism*.

in time. Our experience of a temporal process all at once (without which we could never really experience temporal process at all) establishes the possibility of this. For notions of temporal relations which do not allow for a whole within which different times exist are ultimately as incoherent as are notions of space which do not make space a kind of whole.

If this is correct, all that is is contained in a single divine consciousness within which an inconceivably vast number of streams of finite experience interact and interweave. When the lower level streams of experience which correspond to the basic items postulated in physics enter into appropriately complex relations with each other they form aggregates (and aggregates of aggregates) which are what living things are in themselves, and which underpin the emergence of the streams of consciousness of animals and men. Within such streams of consciousness, more particularly the human, a not self aspect, which is primarily the physical world as it is for us, confronts a self aspect, and serves as its representation of the system of interweaving streams of experience in the midst of which it exists and with which it must interact appropriately in order to survive, communicate with other similar selves, and realize its personal essence as fully as it can.

§6

Thus we are less separate from one another than we seem. What ethical significance does this have? Should we agree with Schopenhauer (who, however, would deny that the unity of the world was that of a single *consciousness*) that the good man, who does not draw the ordinary degree of distinction between himself and others, is piercing through an illusion and living more in the light of the truth than the rest of us?

> If plurality and separateness belong only to the *phenomenon*, and if it is one and the same essence that manifests itself in all living things, then that conception that abolishes the difference between ego and non-ego is not erroneous; but, on the contrary, the opposite conception must be. We find also that this latter conception is described by the Hindus as *Maya*, i.e.,

> illusion, deception, phantasm, mirage. It is the former view which we found to be the basis of the phenomenon of compassion; in fact, compassion is the proper expression of that view. Accordingly, it would be the metaphysical basis of ethics and consist in *one* individual's again recognizing in *another* his own self, his own true inner nature.

It is not only the postulation of a universal consciousness which suggests that our personal distinctness is only partial. We have seen that what each of us essentially *is* is a personal essence, a particular way of cognizing and dealing with the world, which largely dominates our consciousness. But this is not the only essence that does so, for there is also the more generic essence of simply human consciousness, the distinctively human way of experiencing the world, of which it is one specification. Now it is in a manner up to the successive moments in each stream of consciousness, as each takes on the task of choice for its brief moment of existence, whether it regards itself primarily as an instance of one particular personal essence, or of the essence of human consciousness in general. (And there may also be intermediate essences, national or cultural, which some moments of consciousness experience themselves as actualising more significantly than they do either a personal self or humanity, though it is open to much debate how desirable this is and how fully individual in their nature these essences really are.) Whatever choice it actually makes, there is as truly something which can be called Humanity which lives in each moment of human consciousness as there is something called, in my case, Timothy Sprigge, which lives in each moment of my consciousness. Each is a genuine concrete universal. It is a form partially actualised, and pushing towards a fuller actualisation, in a series of instances in each of which its presence is due to or will produce its presence in another. For just as the torch of promoting the Timothy Sprigge way of experiencing the world is handed on from moment to moment of my conscious life, so the task of experiencing and developing the human way of experiencing the world is handed about from one moment of human experience to another, though in a more criss crossing fashion than the merely serial order of those experiences which are distinctively mine. Thus when another suffers it is as true a way of looking at the situation for my consciousness to say to itself: 'That is the very same humanity which suffers in that person as lives in me' as to say 'That is quite another being who is suffering'. Although both ways of looking at the situation are correct, the personal way of looking at it tends to go with a failure to realize the validity of the other way of looking at it, and to that extent is positively erroneous. As we shall see, it would be equally wrong not to realize that it is sheer consciousness as such which lives in every conscious being, whether human or animal, or in the end in every ultimate unit in nature.

Thus although personal identity is, in its way, quite genuine, it becomes an illusion when it is taken as an essentially deeper identity than pertains to that of a reality like humanity as a whole. One aspect of this illusion is the idea that there is a difference between the fact that one person has a certain experience and that another does which is not a matter of a difference in the quality of the experience. One tends to think that if the total nature of another person's experience could be specified there would be some further fact to the effect that it was he who had the experience rather than I.* In truth, however, the fact that he had the experience is one with the fact that the experience had a certain character determined by his particular personal essence. I, Timothy Sprigge, could not have had the experiences of Julius Caesar because what makes them his experiences is that they are specifications of the particular way in which a very individual sort of forceful Roman personality encountered an environment of a particular historical character.

If one finds this hard to accept, and feels that after all it could have been oneself who had those experiences, this is because what is thinking through one when one has those thoughts is not so much a particular person, such as Timothy Sprigge, as humanity, which, in this instance of it, recognizes its capacity to be specified in the Julius Caesar way as well as in the Timothy Sprigge way. In so far as what is calling itself 'I' upon my tongue is humanity, then it is true that I could have had those experiences, the final proof of this being that I did. Thus, for there to be full recognition of the reality of another's suffering,

* What I say here has some similarity to a a theme basic to the work of Derek Parfit.

humanity in me must recognize itself as present equally in him.

Even if the personal essences of two individuals were basically the same, considered apart from the distinct streams of consciousness in which they were being realized, each cross section of that stream would be so conditioned by what went before that it could not have occurred in the other stream.

Thus we can agree with Schopenhauer that moral badness comes from taking the distinction between ourselves and others as having a degree of reality much greater than it possesses, while not dismissing personal identity as a complete delusion from the point of view of ultimate truth. We are all elements in one universal consciousness, and the boundaries between us are in no way absolute. Each moment of human consciousness is both that of a distinct human person and of humanity or indeed the cosmos at large and to the extent that the sense of either of these facts is missing in it it is deluded.

§7

Another implication of these ideas is that we should not confine our view of values to the human, or even animal, world. Certainly animal consciousness (in some appropriate sense of 'animal') is as much a genuine concrete universal as humanity at large or each one of us individually. For there is, surely, a common way of experiencing the world seeking to enhance itself within the consciousness of all creatures with a brain and sense organs at all akin to ours. The most striking sign of the presence of this concrete universal is a pair of eyes. All creatures who negotiate with their environment guided by eyes or other sense organs by which they locate things in a three dimensional space share a very significant common form of consciousness. It is only as it lives in the consciousness of humans that this animal form of consciousness has taken on the form of discursive thought, but having done so it can recognize itself as existing inarticulately in other creatures. The sense of affinity one may have with a horse is not an illusion, nor a matter of some vague similarity. There is something one and the same living within each of us and every horse as truly as there is in you at different stages of your life. In the course of evolution animal consciousness has developed into many different forms, but the common essence remains, as does the dynamics of mutual influence which links all its instances into a single life.

If the panpsychic view of things I have adumbrated is true then we can think of animal consciousness as itself simply a special turn which the concrete universal of sheer consciousness or feeling present in all things has taken with the evolution of animals. That being so, each moment of my conscious life is an instance of sheer feeling, risen to a high degree of articulateness, and as such it can look at the things around it and know that within them this same reality of feeling lives as lives in it. In most cases, however, it seems we can know little of what the more concrete character of that feeling is. Still, we are continually in interaction with the environment – and with all the processes within our own bodies – and this must really consist in the fact that our consciousness is continually influenced by and influencing the multiplicity of forms of feeling which are what environment and world are in themselves. When life is healthy and zestful, or when a sense of union with a greater whole stirs in us, that exchange of influence is wholesome on our side, and just possibly so on the other side too.

Value can only be present when feeling is present but feeling is present everywhere, and all this feeling belongs to one vast unitary experience which the cosmos has of itself. This, it seems clear, is no vast chaotic sensorium of dumb feeling but a rationally organized system, and this seems to imply that it is in some sense self aware and intelligent. Not that we should think of it as carrying on its own individual verbal discourse with itself (though it contains all verbal thinking) but rather as being articulate in some fuller way. Only thus, I believe, can we make sense of the laws of nature (ultimately perhaps a matter of the essences under which the diverse streams of experiential process are conceived by it) not merely as universal generalisations, but as the system in terms of which the universe experiences itself as organized.

Thus the rather narrow conception of the locus of value to which its identification with pleasure and pain seemed to lead is not finally valid. Yet we know so little of the kinds of heightening and lowering of value which go on in the world other than in relation to humans and animals that we can hardly take much account of them. By and large the values in the environment we know about, and can do much to promote, must be those which pertain to the version of it which appears in the not self aspect of our consciousness.

In recent years what is known as an ecological conception of ethics has developed. Humans are condemned for thinking that their practices should be judged only for their bearings on human life. Nor, it is often thought, is this adequately put right merely by extending our concern to individual animals. Rather, man should see himself as but one component in life on this planet, and in the light of this hold 'that a thing is right when it tends to promote the integrity, stability and beauty of the biotic community . . . wrong when it tends otherwise'. (LEOPOLD pp. 224 et ff.)

The desirability of judging man's activities for their bearings on the biotic community, that is, on the integrated system of life on this planet, can be argued either on so-called shallow or on so-called deep ecological grounds. (ARNE NAESS and Part One of SCHERER AND ATTIG.) The shallow ecologist is ultimately only concerned with human welfare, but insists that humans have ignored the extent to which they can only survive prosperously and pleasantly if they are wise and informed guardians of life on the planet. If we ignore the natural underpinning of our lives and ruin it, human life will not survive in any pleasant form. However, the deep ecologist insists that we should not just approach the larger world of nature in terms of its utility for us, but should see the survival of a rich plurality of life forms in thriving ecological relationship as something good in itself quite apart from us. Philosophers of this stamp have also insisted that individual natural realities, great sweeps of wilderness or wild and rushing rivers, can be of intrinsic value, and that we should have a direct concern with promoting the values there. (See REGAN Chapter IX.)

Unfortunately this approach is usually not associated with much reflection on the relation between the world as it is for us and the world as it is apart from its presentation to the human mind. Surely it requires some view as to how much of that which seems valuable to us in nature pertains to it only as object for us rather than as something existing independently in itself.

However, even if much of what is valuable in nature is the value of something which can only exist as object of our contemplation, that does not mean that its value lies merely in its utility for us or even in the enjoyment it gives us. Maybe it can only exist as an object for a consciousness like ours, but that does not mean that its value, when it does exist, lies in something it does for us, considered as something over and against it. The music of Beethoven, or a building or a painting as an aesthetic object, can only really exist for a human consciousness, but they are objects nonetheless with their own kind of reality and value. They are not the individual sensations of any one person, but common objects which rather impart a value to the consciousness in which they appear than derive their value from anything properly called their effects upon that consciousness. Consciousness, of our kind, divides into self and not self, and not self has its own value; it is not merely good – or bad – for what it does for self. Thus in regarding nature, or more specifically nature on this planet, or certain aspects of nature, as intrinsically good, we are acknowledging the value of something not ourselves, even if it is something which can only exist in a consciousness like ours.

But can there not be values in nature in its independent being? Certainly there can be and presumably are, if nature consists of flows of experience in systematic relation. But how can we know what there is in nature which has this independent value? Up to a point, perhaps, we can think that the exemplification of some of the structures specified in science may be especially good, for example, the structural aspects of the inter-relations of species as described by the ecologist, these often being what the deep ecologist especially values. However, it is difficult to see how there can be much value in these structures until they are felt synthetically in one consciousness. The consciousness of most animals cannot fill this role. Only human consciousness, or the unitary consciousness of the universe, can do so, and it is hard to see how man can judge what is better or worse in this way for the absolute consciousness except in terms of what presents the richer spectacle for him. But cannot our direct participation in nature, the colours, the scents, the wind on our skins, give us a sense of the intrinsic value of certain places? Yes, perhaps we may suppose that certain sorts of place are good to be in for us, because there is a certain independent goodness there. However, here again we can only really take the goodness of nature as presented to us as a possible indication of a goodness which is there independently of us. Thus in the end, our only practicable concern must be with nature as it appears to us. This includes, however, the goodness of nature in those aspects which give us a sense of union with a great spiritual whole to which we belong, a sense which may sometimes go somewhere to quench our petty egoisms, and which metaphysical reasoning can (I believe) show to be no mere illusion.

The 'us' in question, however, should not be the merely human. Not only fellow humans, but also animals, as presented to our sight, are signs of distinct worlds of feeling. We can know at least something of what is good and bad within those worlds and recognize it as something of which we should take account as we could not help doing if we were confronted with it more fully. And we may sense that some of the higher animals have forms of life which make a distinctive contribution to the value present in the world as a whole and should be cherished just as a variety of different human forms of life should be.

But is the world really the better for a variety of values? Provided there is a suitable amount of variety within each consciousness, what does it matter whether there is variety in the world at large? Well, if there is a unitary consciousness of the world at large then we can suppose that that is the richer for the variety of types of value within the world. Not that we need think of this as only existing at that rather remote level for something of the richness of the whole filters into each of us.

In this and other ways the notion of a unitary consciousness gives a sense to the idea that it matters what happens in the world as a whole which seems otherwise lacking. Despite our conclusions about the sense of comparing the value of totalities of experiences not all present in one consciousness something puzzling remains as to how two pains can be worse than one when they are not felt as a unity and how a variety of different ways of enjoying the world forms a better totality than a monocultural situation. However, if there really is a universal experience of which our experiences are all aspects they can add up therein to make larger felt unities than those of any one person's consciousness.

It is true that the unifications of our different streams of consciousness in the absolute consciousness are inaccessible to finite individuals like us and this may seem to make their value ethically irrelevant. But there is something of very great concern to us of which these postulated unities must be supposed to be the ground, namely that each phase of finite consciousness – while a distinct reality in its own right – comes to itself with a sense, in its broad character infallible, of the history of the self to which it belongs and of the community to which that self belongs. Our experiences interpenetrate. 'No man is an island.' The universal consciousness (as I see it) is the medium of interpenetration, in virtue of which a developed consciousness, seeking its own self realization, cannot do this in detachment from concern with the good of those among whom he lives.

However, we cannot exist as persons, if we live always at the lofty level of the sense of identity with all around us. Schopenhauer is right that the compassion, which is the world's best hope for alleviation of the evil which some fatality has built into it, turns on an insight into an underlying identity between all conscious beings. But ethics is conceived too narrowly if we make such compassion its sole basis. We must listen also to those, like Spinoza, acccording to whom codes of conduct can only be recommended intelligently to beings who see them as, by and large, enriching their personal lives. Some of the time these two bases for a reasonable code of behaviour may come in conflict, in which case rather than plump for one or the other we must seek some *modus vivendi*. But for the most part, when each is thought out intelligently, they are mutually reinforcing.

Bibliography

This is mainly a list of sources for positions ascribed to authors in the text. In the absence of a more specific reference all works listed here by an author are relevant. Names in capitals and in square brackets in the text refer to the only or first book by an author in this list unless a numeral N indicates that it is the Nth one listed.

The list below includes names of works by a few authors not otherwise referred to which are closely related to matters discussed in the text.

Attfield, Robin: *The Ethics of Environmental Concern* (Basil Blackwell, Oxford, 1983) Chapter VI.

Ayer, A.J.: *Language, Truth and Logic* (London, 1946).

——: *Philosophical Essays* (London, 1954) Chapter 10.

Bambrough, Renford: *Moral Scepticism and Moral Knowledge* (Routledge & Kegan Paul, London, 1979).

Bayles, Michael (ed.): *Contemporary Utilitarianism* (Anchor Books, New York, 1968).

——: *Ethics and Population* (Schenkman Publishing Company, Cambridge, Mass. 1976).

Bentham, Jeremy: *An Introduction to the Principles of Morals and Legislation* (first published 1789; various editions).

Blackburn, Simon: *Spreading the Word* (Clarendon Press, Oxford, 1984) Chapter 6.

Bradley, F.H.: *Ethical Studies* 2nd Edition, with additions (Clarendon Press, Oxford, 1927).

Brentano, Franz: *The Origin of Our Knowledge of Right and Wrong* (first published 1889. English translation edited by Oskar Kraus, Routledge & Kegan Paul, London, 1969).

Broad, C.D.: *Five Types of Ethical Theory* (London, 1930).

Butler, Joseph: *Fifteen Sermons preached at the Rolls Chapel*, London, 1726. (The five most philosophical sermons are republished, together with an appendix from Butler's *The Analogy of Religion* in Joseph Butler: *Five Sermons and a Dissertation Upon Virtue*, Bobbs-Merrill, Indianapolis, 1950.)

Cooper, Neil: *The Diversity of Moral Thinking* (Oxford, 1981).
Dawkins, Richard: *The Selfish Gene* (Oxford University Press, 1976).
Deci, D.L.: *Intrinsic Motivation* (New York, 1975).
Dworkin, Ronald: *Taking Rights Seriously* (Duckworth, London, 1977).
——: *Law's Empire* (Fontana Press, London, 1986).
Edwards, Rem B.: *Pleasures and Pains, A Theory of Qualitative Hedonism* (Cornell University Press, 1979).
Epicurus: *Letters, Principal Doctrines, and Vatican Sayings* (transl. Russell M. Gear, Bobbs-Merrill, Indianapolis, 1964).
Ewing, A.C.: *The Definition of Good* (New York, 1947).
Findlay, J.N.: *Meinong's Theory of Objects and Values* (2nd edition, Oxford 1963).
Foot, Philippa: *Virtues and Vices* (Oxford: Blackwell, 1978).
Foot, Philippa (ed.): *Theories of Ethics* (London, 1968).
Frankena, W.K.: 'The Naturalistic Fallacy' in Foot (ed.).
Glover, Jonathan: *Causing Death and Saving Lives* (Penguin Books, 1977).
Gosling, J.C.B.: *Pleasure and Desire: The Case for Hedonism Reviewed* (Oxford, 1969).
Grayling, A.C.: *The Refutation of Scepticism* (Duckworth, 1985).
Haksar, Vinit: *Equality, Liberty and Perfectionism* (Clarendon Press, Oxford, 1979).
Hare, R.M.: *The Language of Morals* (Clarendon Press, Oxford, 1952).
——: *Freedom and Reason* (Clarendon Press, Oxford, 1963).
——: *Moral Thinking* (Clarendon Press, Oxford, 1981).
Harmann, G.: *The Nature of Morality* (Oxford University Press, 1977).
Harrison, Ross: *Bentham* (Routledge & Kegan Paul, 1983).
Hobbes, Thomas: *Leviathan* (first published 1651; many modern editions).
Honderich, Ted (ed.): *Morality and Objectivity* (Routledge & Kegan Paul, London, 1985).
Hume, David: *A Treatise of Human Nature* (first published 1739–40; Oxford, 1888, ed. L.A.Selby-Bigge: numerous reprints).
——: *Enquiries concerning the Human Understanding and concerning the Principles of Morals* (first published 1748 and 1751; Oxford, 1902, ed. L.A.Selby-Bigge; numerous reprints).
Husserl, Edmund: *Ideas: General Introduction to Pure Phenomenology* (transl. W.R. Boyce Gibson, Allen & Unwin, London, 1931).
Hutcheson, Francis:
——: *Inquiry concerning Moral Good and Evil* (1725).
——: *Illustrations upon the Moral Sense* (1728).
There is no modern edition of these works and most readers will need to turn to the selections in Selby-Bigge below.
Kant, Immanuel: *Groundwork of the Metaphysic of Morals* (A convenient translation and edition is: *The Moral Law* or *Kant's Groundwork of the Metaphysic of Morals* transl. and ed. by H.J. Paton, Hutchinson, London, 1948.)
Leopold, Aldo: *A Sand County Almanac* (first published 1949: Oxford University Press, London, 1966).

Lewis, C.I.: *An Analysis of Knowledge and Valuation* (Open Court, 1946).
Lovibond, Sabina: *Realism and Imagination in Ethics* (Blackwell, Oxford, 1983).
Lyons, David: *Forms and Limits of Utilitarianism* (Clarendon Press, Oxford, 1965).
Macintyre, Alasdair: *After Virtue: A Study in Moral Theory* (Duckworth, London, 1981).
Mackie, J.L.: *Ethics, Inventing Right and Wrong* (Penguin Books, 1977).
Manser, Anthony and Stock, Guy: *The Philosophy of F.H. Bradley* (Clarendon Press, Oxford, 1984).
McDowell, John: 'Are Moral Requirements Hypothetical Imperatives?' (in *Proceedings of the Aristotelian Society, Supplementary Volume, 1978*).
——: 'Aesthetic value, objectivity, and the fabric of the world' (in *Pleasure, Preference and Value*, ed. Eva Schaper, Cambridge University Press, 1983).
——: 'Virtue and Reason' (in *The Monist*, July 1979).
Midgley, Mary: *Wickedness* (Routledge & Kegan Paul, London, 1984).
Mill, John Stuart: *Utilitarianism* (first published, 1863; numerous modern editions).
Moore, G.E.: *Principia Ethica* (Cambridge University Press, 1903).
——: *Ethics* (Home University Library, Oxford University Press, London, 1912).
——: 'The Conception of Intrinsic Value' in *Philosophical Studies* (Routledge & Kegan Paul, London, 1922).
Naess, Arne: 'The Shallow and the Deep, Long-Range Ecological Movement' (in *Inquiry* 1973).
Nagel, Thomas: *The Possibility of Altruism* (Clarendon Press, Oxford, 1970).
——: *Mortal Questions* (Cambridge University Press, London, 1979).
——: *The View from Nowhere* (Oxford University Press, 1986).
Norman, Richard: *The Moral Philosophers* (Oxford University Press, London, 1983).
Parfit, Derek: *Reasons and Persons* (Clarendon Press, Oxford, 1984).
Price, Richard: *A Review of the Principal Questions in Morals* (1st edn 1758; reprinted Oxford, 1948 with introduction by D.D. Raphael).
Rawls, John: *A Theory of Justice* (Oxford University Press, London, 1972).
Regan, Tom: *All that therein Dwell* (University of California Press, 1982).
——: *The Case for Animal Rights* (Routledge & Kegan Paul, London, 1983).
Ross, W.D.: *The Right and the Good* (Clarendon Press, Oxford, 1932).
——: *The Foundations of Ethics* (Clarendon Press, Oxford, 1939).
Royce, Josiah: *The Religious Aspect of Philosophy* (first published 1885. Republished Peter Smith, Gloucester, Mass. 1965).
——: *The Philosophy of Loyalty* (first published 1908. Republished

Hafner Publishing Company, New York, 1971).
Ryle, Gilbert: *Dilemmas* (Cambridge University Press, Cambridge, 1962).
Santayana, George: *The Sense of Beauty* (first publ. 1896; Dover Publications, New York, 1965).
——: *Reason in Science* (Charles Scribner's Sons, New York, 1906).
——: *Winds of Doctrine* (Dent, London, 1913).
——: *Realms of Being* (Charles Scribner's Sons, New York, 1942).
——: *Dominations and Powers* (Charles Scribner's Sons, New York, 1951).
Sartre, Jean-Paul: *Existentialism and Humanism* (first published 1946; English transl. by Philip Mairet, Methuen, London, 1948).
Scheffler, Samuel: *The Rejection of Consequentialism* (Clarendon Press, Oxford, 1982).
Scheler, Max: *The Nature of Sympathy* (first published 1913: transl. Peter Heath, Routledge & Kegan Paul, London, 1954).
Scherer, Donald and Attig, Thomas: *Ethics and the Environment* (Prentice-Hall, New Jersey, 1983).
Schilpp, P.A.S. (ed.): *The Philosophy of G.E.Moore* (New York, 1952).
Schlick, Moritz: *Problems of Ethics* (transl. David Rynin of *Fragen der Ethik*, first published 1930: Dover Publications, New York, 1962).
Schopenhauer, Arthur: *On the Basis of Morality* (first published in 1841; transl. E.F.J. Payne, Bobbs-Merrill, Indianapolis, 1965).
——: *Essay on the Freedom of the Will* (first published in 1841; transl. Konstantin Kolenda, Bobbs-Merrill, Indianapolis, 1960).
Selby-Bigge, L.A.: *The British Moralists*, Vols 1 and 2 (Dover Books, New York, 1965, which is precise reproduction of original Clarendon Press, Oxford edition of 1897).
[This contains substantial parts of the work of Hutcheson (including the whole of his *Inquiry concerning the original of our Ideas of Virtue and Moral Good*), Richard Price, Butler, Adam Smith, and other important seventeenth- and eighteenth-century moral philosophers]
Sidgwick, Henry: *The Methods of Ethics* (Macmillan, London, 1st edn, 1874, 7th edn, 1907, and often reprinted).
Singer, Peter: *Animal Liberation* (Jonathan Cape, London, 1976).
——: *Practical Ethics* (Cambridge University Press, Cambridge, 1979).
Skinner, B.F.: *Science and Human Behaviour* (Free Press, New York, 1953).
Smart, J.J.C.: *An Outline of a System of Utilitarian Ethics* (in *Utilitarianism, For and Against* by J.J.C. Smart and Bernard Williams, Cambridge University Press, Cambridge, 1973).
Smith, Adam: *The Theory of Moral Sentiments* (first publ. 1759; Hafner New York, 1948).
Sperry, Roger: *Science and Moral Priority* (Basil Blackwell, Oxford, 1983).
Spinoza, Baruch: *The Ethics* (first published 1677). (There are various translations. That by Samuel Shirley, Hackett Publishing Company, 1982 is recommended, but the one in *Spinoza Selections*, Charles

Scribner's Sons, 1920, and often reprinted, will serve.)
Sprigge, T.L.S.: *Santayana: An Examination of his Philosophy* (Routledge & Kegan Paul, London, 1974).
——: *The Vindication of Absolute Idealism* (Edinburgh University Press,1983).
——: *Theories of Existence* (Pelican Books, 1984).
——: The following articles (especially 'Punishment and Responsibility') supplement views presented in this book:
——: "Definition of a Moral Judgement" (*Philosophy*, 1964). (Reprinted in *The Concept of Morality* ed. Wallace and Walker, Methuen, London, 1973).
——: 'A Utilitarian Reply to Dr McCloskey' (*Inquiry*, 1965) (reprinted in *Contemporary Utilitarianism*, ed. M. Bayle, Anchor Books, New York, 1965, and, in shortened form, in *Philosophical Perspectives on Punishment*, ed. G.Ezorsky, State University of New York Press, Albany, 1972).
——: 'Punishment and Responsibility' in *Punishment and Human Rights*, ed. Milton Goldinger (Schenkmann Publishing Co., Mass. 1974).
——: 'Metaphysics, Physicalism and Animal Rights' (*Inquiry*, 1979).
——: 'Non-Human Rights: An Idealist Perspective' (*Inquiry*, 1984).
——: 'Idealism and Utilitarianism: A Rapprochement' (*Philosophy*, October 1985).
——: 'Are there intrinsic values in nature?' (in *Journal for Applied Philosophy* Vol 1, 1987).
Stevenson, Charles L.: *Ethics and Language* (Yale University Press, New Haven, 1944).
——: *Facts and Values* (Yale University Press, New Haven, 1963).
Taylor, Charles: *The Explanation of Behaviour* (Routledge & Kegan Paul, London, 1964).
Taylor, A.E.: *The Elements of Metaphysics* (Methuen, London, 1903 and 1961).
Toulmin, Stephen: *The Place of Reason in Ethics* (Cambridge University Press, 1953).
Waldron, Jeremy (ed.): *Theories of Rights* (Oxford University Press, London, 1984).
Whitehead, A.N.: *Process and Reality* (Corrected Edition, The Free Press, New York, 1979).
Williams, Bernard: *Ethics and the Limits of Philosophy* (Fontana Press, London, 1985).
——: *A Critique of Utilitarianism*, in *Utilitarianism, For and Against* (eds J.J.C. Smart and Bernard Williams, Cambridge University Press, Cambridge, 1973).
——: *Morality* (Cambridge University Press, Cambridge, 1972).
Young, P.T.: *Emotion in Man and Animal* (John Wiley, New York, 1943).
——: *Motivation and Emotion* (John Wiley, New York, 1961).

Index